OPEN HOUSE

A Home Inspection Guide for Buyers & Sellers

Bryan Trandem

Table of Contents

CREATIVE PUBLISHING international

Copyright© 2003
Creative Publishing
International, Inc.
18705 Lake Drive East
Chanhassen, MN 55317
1-800-328-3895
www.creativepub.com
All rights reserved
Printed on American Paper by:
Quebecor World

10 9 8 7 6 5 4 3 2 1

President/CEO:
  Michael Eleftheriou
Vice President/Publisher:
  Linda Ball
Vice President/Retail Sales
  & Marketing: Kevin Haas

Executive Editor: Bryan Trandem
Creative Director: Tim Himsel
Managing Editor:
  Michelle Skudlarek
Editorial Director: Jerri Farris
Author: Bryan Trandem
Additional Writers & Editors:
  Andrew Karre, Karen Ruth,
  Dane Smith

Copy Editor:
Shannon Zemlicka
Lead Art Director: Bill Nelson
Illustrator: Bill Nelson
Stock Photo Editor:
Julie Caruso
Director, Production Services &
Photography: Kim Gerber
Production Manager:
Helga Thielen
Studio Services Manager:
Jeanette Moss McCurdy
Photo Team Leader:
Tate Carlson
Photographers: Bill Nelson,
Bryan Trandem
Tate Carlson

OPEN HOUSE
Created by:
The Editors of Creative Publishing
international, Inc.

Library of Congress Cataloging-in-
Publication Data

Trandem, Bryan
Open house: A home inspection
guide for buyers and sellers /
Bryan Trandem
p.cm
Includes index
ISBN 1-58923-060-4 (soft cover)
1. House buying. 2. House selling.
3. Real estate business. I. Title

HD1379.T72 2003
333.33'83--dc21
2003040979

Chapter 1

Open for Viewing
How to Use This Book

Buying a new home is one of the most dramatic events of your life. In terms of life changes, it ranks right up there with getting married or having kids. For most of us, a home represents the single biggest investment we'll ever make, and so we very much want to get it right. Each of us is looking for a dream home that will safely shelter our family, but we're also pretty darned terrified of finding ourselves saddled with a money pit that will cost us peace of mind, not to mention our life savings.

Like all major life events, shopping for a home can be full of stress—it's exciting, nerve-racking, exhausting, and invigorating, all at the same time. No event is bigger in terms of financial investment, and only a few events have greater impact on our lives. And selling a home can be just as stressful as buying one.

I've written this book for both types of readers: those shopping for a new home, and those preparing one for sale. After all, the two go hand-in-hand, making it likely you're doing both things simultaneously—selling your existing home and getting ready to move to another one.

Whenever market conditions become competitive, homes sometimes get sold within days or even hours of their appearance on the market. For this reason, there's often simply no time for a slow, methodical evaluation before you make a bid; nor is there time to call in a professional inspector to do an in-depth study of the home before you make an offer. You often need to move very quickly, so you'll need to know what to look for from the very moment you park in front of the house. "Pressure from sellers, realtors, and other home buyers to make a quick purchase decision is making it even more important than ever to be an educated home buyer who knows what to look for when shopping for a resale home," says Kenneth T. Austin, president of HouseMaster. This is one reason why this book can be so useful.

I've been writing and editing books on home improvement and landscaping for almost 20 years, and in that time I've learned a

Like all positive life events, shopping for a home can be full of stress—it's exciting, nerve-racking, exhausting, invigorating, all at the same time.

great deal about what can go wrong with a home and how to fix those problems. Friends, colleagues, relatives, and neighbors often come to me for opinions on their home problems, as well as advice when they're preparing to buy or sell a home. Even though I've been in the same home for the last 10 years or so, Sunday afternoons often find me visiting real estate open houses or model home showings—sometimes to help friends who are house-hunting, but often just for the sheer fun of it.

I've found that a small handful of tools and materials can be very useful when you're looking at homes with a serious interest in buying. They won't cost you all that much, and you probably already own many of them (right):

- Pencil and notepad
- Flashlight
- Electrical circuit testers
- Magnet (to identify iron pipes)
- Small ball or large marble (to check floors for level)
- Torpedo level
- Tape measure
- Compass (to determine exposure of house)
- Awl
- Small binoculars (to inspect roof and vents)
- Magnifying glass
- A small digital camera (ask before taking photos)

Whenever market conditions become competitive, homes sometimes get sold within days or even hours of their appearance on the market.

A small handful of tools and materials can be very useful when you're evaluating a house.

Because the subject of health risks in the home is so important, we've covered this in its own chapter, where we'll show some of the testing kits that are useful.

Using This Book

Buying a home. If you're shopping for a home, use this book as a guide for inspecting homes you might want to buy. I've organized the book in an easy-to-follow fashion, from researching and evaluating neighborhoods, to systematic walkthroughs of every element of a house's structure and mechanical systems—landscape to wallpaper, basement to attic.

Going in, I advise you to be realistic. Logic tells you that it's really not possible to buy a home that is absolutely free of all problems. You'll quickly grow discouraged if this is your aim. Even newly-built homes, if you look very closely, have trouble areas. And at first glance, older homes can seem to be a morass of trouble spots. The good news is that many of these issues are largely cosmetic and not all that serious. That's what this book is about—looking past the cosmetic issues and past the obvious, glaring problems that everyone can see to identify those that are more likely to escape notice. Peeling wallpaper is obvious to everyone, but the signs of inadequate plumbing are not.

Statistics tell us that 40%—two out of every five homes that you'll visit while shopping for a new home—have at least one big problem that could cost you as much as $15,000 to repair. The goal of this book is to help you spot this kind of problem and evaluate the house fairly.

For efficiency's sake, *Open House* is organized as though you're doing your inspection on a single pass. It's entirely possible to do it this way, but it's more likely that you're going to make two, three, or even four visits to the house, gradually getting more in-depth with the level of detail you look for. The order I've presented here is the way I like to inspect a home, but don't feel compelled to do it my way. It's fine to jump around in this book in whatever way suits your method.

Selling a home. If you're selling your home, you can use this book to anticipate the types of problems that prospective buyers might be looking for. Many problems are relatively easy to fix, and making these repairs can make your home much more appealing to buyers. I've included *If You're Selling* sidebars especially for you. These tips will help you prepare for the types of questions prospective buyers will ask, and give you ways to position your home in the best possible light.

Above all, I encourage you to be honest about the history and condition of your home when you're trying to sell it. A prospective buyer will be impressed if you're honest about the flaws, which he or she will probably find anyway, but will look elsewhere in a hurry if you're caught hiding a significant problem.

Working with professional inspectors. By no means do I intend for this book to take the place of a formal home inspection by a licensed home inspector. Once you zero in on the house you want to buy and make a bid, it's a very good idea to hire a home inspector to make an in-depth evaluation of the home. For the cost of a few hundred dollars—Uusually between $200 and $400—an inspector will dig into fine details and reassure you that the home is solid.

It's beyond the scope of any book to pinpoint every possible

At first glance, older homes can seem to be a morass of trouble spots. The good news is that many of the issues are largely cosmetic and not all that serious.

problem with a home. What I can do, though, is show you what to look for, and how to spot less-than-obvious signs of problems that could seriously affect your quality of life. In this way, you can maximize the chances that the professional inspector will give the green light to the home you've chosen.

Each chapter of *Open House* includes an Inspection Checklist feature. It's a good idea to photocopy these pages and use them as worksheets when doing your inspection of a home.

Other Features

Deal Killers will help you identify some problems that might be so serious or expensive enough to cause you to reject the home altogether. At the very least, identifying expensive problems may help you negotiate a more reasonable buying price. If you are an energetic and skilled do-it-yourselfer, you might even prefer to look for a fixer-upper as a target for your sweat-equity investment.

Signs of Gold. No one likes to focus entirely on the negative, so in most chapters, we'll show you some things to look for that will reassure you that the house has been well cared for and might be a great buy.

Tracks of the Poor Craftsman. Shoddy workmanship can be the result of a poor contractor or a do-it-yourself homeowner who has exceeded his or her abilities. I'll show you some of the warning signs that a poor craftsman has been at work

If You're Selling boxes give tips on preparing your home for sale. You'll get advice on the best home improvements to make— those that give the most bang for the buck when it comes to resale value, and how to present your home to the best advantage during open houses.

Thank You...

Among the people who deserve special thanks are my colleagues Dane Smith, Karen Ruth, and Andrew Karre, who helped draft several chapters; and Jerri Farris, who challenged my language and edited the work with her usual incomparable skill.

Among the people who shared their homes and experiences with us: Aimee Jackson, Jon Simpson, Karen Kreiger, Jerri Farris, Rhonda Freeman, John & Lea Dahl, Jim & Karen Touchi-Peters, John & Mary Foley, Kaye Butler, Tom Kalgren, Bill Nelson, Jeanette Moss McCurdy, Serena McDurmott, Michelle Skudlarek, Michael Eleftheriou, John Hudgens, and Hillary Green.

Chapter 2

Let the Search Begin
Finding Good Houses in Good Neighborhoods

The home you purchase will maintain or increase its value in a pleasant, well-kept neighborhood.

Take a deep breath. You're about to launch into one of life's great adventures—choosing a home for yourself and your family. But don't begin calling real estate agents and going through the want ads just yet. The first step is not the one you might be expecting.

Finding a good place to live isn't just about finding a well-built, well-maintained house that happens to be for sale. It's also about finding a neighborhood and community that are pleasant and safe, and that provide the services and facilities you need and want.

In fact, I've always found the quality of the neighborhood to be almost as important as the house itself. You can always improve a substandard house in a great neighborhood—in fact, doing just this is a wonderful investment. But it's much harder to have a meaningful impact on a questionable neighborhood. A great house in a bad part of town is no bargain in the long run, so let's talk first about how to zero in on promising neighborhoods.

I've always found the quality of the neighborhood to be almost as important as the house itself.

Step One: Finding Promising Neighborhoods

The first rule of thumb: take your time when looking for a neighborhood. Unless there are compelling reasons why you have to buy immediately, it's a very good idea to take as much as a full year to evaluate the various neighborhoods in a moderate- to large-sized city and get a sense of their quality of life. Even if you're just moving across town in a community you know well, you won't regret taking your time to walk the streets of the neighborhood, talk to neighbors, read neighborhood newsletters, watch the changing of the seasons, check on on crime statistics and government records, and so forth. Real estate agents are full of tales about folks who buy hastily, then find themselves moving within a few months.

New to town? What if you're moving from some distance away to a new town, one you don't yet know at all? If practical, it's a fine idea to rent for a while before you commit to purchasing a home. Again, take as much as a full year to learn your city before committing to a purchase. It's actually quite an adventure to learn a new city, and you'll never regret taking this time. When a new colleague and her husband moved to town, they lived in an apartment for sixteen months, gradually learning about the city of Minneapolis and its suburbs. The two of them then made an educated decision and bought a house that was perfectly suited to their needs.

Of course, it isn't always possible to take this much time— perhaps you're just impatient, or maybe there are financial or personal reasons you need to buy quickly. If you find it necessary to buy a home as soon as you move to a new community, try to make several preliminary visits before you move permanently. The more you know about your new city, the more informed your decision will be. If you know anybody at all in your new city, ask for advice on neighborhoods to look into.

It's actually quite an adventure to learn a new city, and you'll never regret taking this time.

The Internet. In recent years, this has become one of the richest sources of information imaginable for real estate details. All the major national real estate chains maintain sites that will help you look for houses on-line, but they can also teach you quite a bit about neighborhoods. You can evaluate neighborhoods based on average home age, average size, and crime statistics. You can even take slide show tours of available houses in the zip code area you specify, or in the price category or feature category you choose.

The Internet can be a lifesaver if you're moving to a community you've never visited before. If you don't have access to the Internet at home or at work,

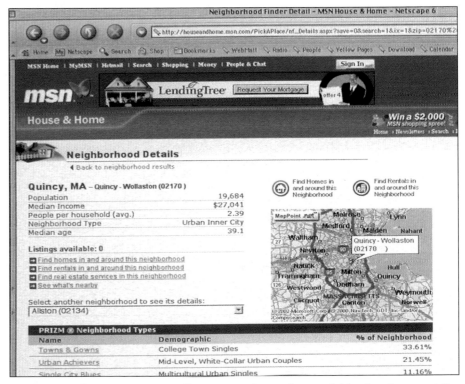

Internet web sites are an excellent way to learn about neighborhoods, especially if you live some distance from the community you're moving to.

virtually all public libraries have computers you can use for on-line browsing. Check the resource guide in the back of this book for a listing of helpful sites.

Newspaper supplements. Most real estate open houses are advertised in the classified section of weekend editions of local newspapers, but there won't be much information there other than the asking price and basic home details—number of bedrooms, baths, etc. You won't find much about the quality of a neighborhood. But newspapers sometimes profile individual neighborhoods in these weekend real estate supplements, and from these articles you can often learn good information about the history and demographics of a neighborhood.

Real estate shoppers and newspaper supplements provide a glimpse of the types of homes available in different neighborhoods.

Real estate shoppers. Grocery markets, convenience stores and other service businesses often have racks that offer free shopper magazines. These shoppers are often localized for the area, and can be a good place to easily find lots of ads. They are usually published by a real estate firm, or groups of real estate firms, so they don't show every home that might be for sale in the area. But they usually include small thumbnail photos of each house being offered, so you can get a sense of whether a neighborhood has home styles that appeal to you.

Real estate agents. It's possible to simply call up a real estate agent to arrange a meeting. The agent will ask about your needs and interests and direct you to neighborhoods and homes that meet your stated needs. Picking an agent you are comfortable with is crucial. A good real estate agent is a gift from the heavens; a poor agent…well, they seem to hail from another place.

A good real estate agent can become a valued acquaintance, even a friend.

Strangely, I've known some agents who seem quite unable to hear what customers have to say and insist on telling them what they want. If you run across one of these, you're better off looking for another agent or choosing to work on your own. A good agent, though, can become a valued acquaintance, even a friend. Remember, however, that you're still fishing for information on neighborhoods. For the time being, resist the agent's suggestion that you schedule a visit to a home.

Random drive-bys. This method is by far my favorite. If you have the luxury of time, few things are more enjoyable than driving through areas of a city or town on a leisurely afternoon when the weather is pleasant and simply looking at the neighborhoods. If you've identified an interesting neighborhood through research, you'll now want explore it in person, if at all possible.

If you spot a house with a "for sale" sign, so much the better. Sometimes the sign on a house may have a box with leaflets that describe the home. Grab one; it should tell you enough for you to know if you want to investigate further. If there's no information sheet, you can call the telephone number on the sign to learn more from the listing agent or homeowner.

Step Two: Exploring the Neighborhood

After you've identified a promising neighborhood through one of the methods I've described, it is enormous fun to simply spend time driving, walking, or biking the streets. And it's a good idea to do this at different times of the week—weekends as well as weekdays—to get an accurate feel for the neighborhood. A neighborhood may have a much different feel on Wednesday afternoon than it does on Saturday night.

Nothing gives you a better feel for a neighborhood than walking or biking its streets. If you find homes posted "for sale" that interest you, you may be able to do a preliminary inspection just by looking at their yards and exteriors. If an interesting home has an alley, drive through it to catch a glimpse of the backyard and garage. If you're feeling gutsy, you might even stop to talk to the owner and ask some preliminary questions about the house.

Spend a few afternoons browsing the neighborhood you've pinpointed. Some things to look for:

Home improvements. Look for signs that neighbors are in the routine of maintaining or improving their homes. Sometimes you might spot building permit notices in windows of homes. Or you might recognize that many of the homes have had tasteful, well-planned structural additions

After you've identified a promising neighborhood, it is enormous fun to simply spend time driving, walking, or biking the streets.

Socially active residents create a neighborhood that's great to live in. Look for neighborhoods that sponsor regular picnics and other get-togethers.

Permit-approved remodeling is a good sign, signaling that owners are investing in their homes and their neighborhood.

A very unusual home nearby makes me smile, but it might make you wince. Either way, such a home makes it easy to give visitors directions to your house.

made. If your general sense is that the neighborhood is on the rise—that its citizens are investing in improving their homes—this can be a very good sign.

Eccentric homes. To some people, it's critical to live in a neighborhood where the appearance of the homes meets traditional expectations, but other people greatly appreciate diversity, even when it strays into eccentricity. It's true, though, that in a neighborhood of a predominant look and style, a single eccentric home can have a dramatic impact on the atmosphere. If this kind of thing is important to you, you'll probably want to survey the neighborhood carefully to make sure no one violates your ideas of what is acceptable.

Parks & recreation areas. Open recreational spaces within walking distance are a good sign in a neighborhood. Experts know that population density can have an enormous impact on the quality of living, so a neighborhood with parks or nature reserves creates a better quality of life. You might want to think twice, though, about living immediately adjacent to a large park that brings in lots of outside traffic.

Traffic patterns. Look at how automobile traffic moves through a neighborhood. Does the neighborhood have lots of commuter traffic, which can make the streets hazardous and unpleasant? Or are there traffic management features, such as speed bumps or intersection islands that make the streets accessible mostly to local traffic? Does the neighborhood have sidewalks and bike lanes for non-motorized traffic? This is usually a good sign.

Churches. A neighborhood church is a mixed blessing. It can certainly represent convenience, if you are a worshiper and the church is of a denomination and style that matches your leaning. And a church generally makes for a quieter, less congested neighborhood—except on days where there are worship services or meetings. Remember, though, that because churches aren't taxed, assessments for street and sewer repairs might turn out to be higher than in neighborhoods where these costs are spread among more residences. In my first neighborhood, our assessment for street repairs was quite steep, because a large church-owned cemetery nearby meant fewer homeowners to carry the costs.

Schools. If you have school-age children, or are planning a family, pay attention to the locations and reputations of the local schools. A good school within walking distance for your children is a very good thing. If you don't have kids, though, a nearby school might be a negative. Living adjacent to a middle school or high school, for example, will almost certainly guarantee that you'll deal with loitering teenagers from time to time.

Proximity to undeveloped areas. If this is important to you, make sure to check with local authorities to make sure that the land is protected against future development.

Mass transit. If you commute to work, look at the neighborhood for access to mass transit—bus stops, freeway entrances, etc. Having easy access to such features can greatly simplify your life— provided they are close, but not too close. Living right smack along a freeway or major highway isn't the most pleasant thing in the world.

Signs. Some signs of a good neighborhood are, well, actual signs. On some blocks, you may see a sign in a window that identifies the home as a safe house for any child who finds himself in danger. Or an alley may have a sign cautioning folks to watch for children. A school or park where a marquee announces a neighborhood event is an indication of a cohesive neighborhood, as are neighborhood watch signs. All these things are positives—indications you've found a neighborhood where the residents look out for one another.

Demographics. It's generally pretty easy to tell if a neighborhood has lots of families with young kids (bikes in driveways, play structures in yards), quiet empty-nesters (carefully

Traffic control features, like this intersection island, can help reduce traffic speeds and make a neighborhood more pleasant to live in.

A neighborhood school can be a great convenience if you have kids, a pleasant feature if you like kids, but a liability if you're annoyed by kids. In one neighborhood I know, the trees spout toilet paper each time the nearby school wins an athletic contest.

A small retail district with stores that offer services you want makes for an appealing neighborhood.

tended gardens, wrought-iron patio furniture), or young singles (hot tubs, sports sedans in driveways). You'll probably enjoy the neighborhood more if there are folks who share your lifestyle and interests, whatever they might be.

Trees. As you'll learn in a later chapter, trees can pose trouble for a home, but in general, plentiful trees make for a better neighborhood. Trees absorb sound and pollution, create pleasant shade, block strong winds, and beautify a neighborhood. I, for one, always gravitate to neighborhoods with healthy, stately trees.

Retail stores. It can be great to have easy access to your favorite type of shopping places. My current home is within four blocks of a coffee shop, a garden center, a movie rental shop, a hardware store, a grocery store, and a library. For me, all that's missing is a bookstore, but six out of seven isn't bad. The point is, pay attention to the types of stores within easy distance of your targeted neighborhood, and score pluses for desirable businesses and minuses for undesirable ones. When I shopped for my first home, I was nearly settled on what appeared to be a very nice working-class neighborhood—until I realized that a small retail zone a few blocks away featured an erotic massage parlor that drew a group of shady visitors. I know people who regret moving into neighborhoods that featured gas stations, liquor stores, or coin-operated laundries.

Step Three: Playing Detective

Some information about a neighborhood won't be made obvious by looking at it. Between visits, get on the telephone or go on-line to learn more about the neighborhood that might contain your future home.

Community newsletters. In general it's a good sign anytime a community can support a regularly published newsletter. This can be a great source of information regarding local services, clubs and organizations, and even crime statistics. A neighborhood where there is an occasional act of graffitti is a much different place than one where rape and robbery are frequently reported.

City Hall. A little digging will almost always produce information or copies of town planning documents. These can tell you if there are major changes planned in the neighborhood, such as zoning alterations, street construction, or utility changes.

Neighborhood newsletters have a wealth of information about the area, including crime reports.

Deal Killers:

If many houses are for sale in a neighborhood, it's wise to ask why. It's possible that this is nothing more than a coincidence, as was the case with the neighborhood shown here, where one family was moving because of a job change, another because they were retiring to another state, and another because of a divorce. But it's also possible there is an underlying reason why families are leaving a neighborhood en masse. It might be because of a planned freeway expansion, the addition of a runway at a local airport, or some environmental health issue. Ask plenty of questions if you see this situation in a neighborhood.

Always investigate the reasons when you see a neighborhood with a number of homes for sale.

Taking Aim

Now that you've selected a few interesting neighborhoods, it's time to find promising houses to inspect.

Finding houses for sale is very easy— real estate is one of the most heavily advertised commodities that exists. The same sources you used to find promising neighborhoods will also provide hundreds of homes that are for sale. The more challenging task is to come up with a list of features that you're looking for in a home. It's surprising how often people fail to give serious thought to this, but if you don't make a list, you might waste a lot of time visiting homes that don't really meet your needs.

The list on page 16 notes some of the features that many people find important—only you can decide which are crucial to you. Include everybody in the family for this discussion, and consider how your needs will change as your family evolves. Your list will undoubtedly have some of the features of the one shown here, as well as others of your own choosing. Have this list in hand when you begin selecting houses to visit.

When it comes time to begin looking at houses, the hardest part can be limiting yourself. I strongly suggest that you not try to visit more than two or three homes on a given weekend. If you can take a few vacation days, plan to visit and inspect one or two homes

Finding houses for sale is very easy—the more challenging task is to come up with a list of features that you're looking for in a home.

HOME FEATURE CHECKLIST

- ❏ **Age of home.** For some folks, a relatively new home is strongly appealing. Others look for older homes that feature craftsmanship that is rare in modern homes.
- ❏ **Finished square footage.** For growing families, bigger is probably better, but if you're empty nesters, a more modest floor space will be easier to maintain.
- ❏ **Architectural style.** Sprawling ranch or quaint bungalow? Tudor stucco or stainless-steel-and-glass modern?
- ❏ **Lot size.** Do you like a big yard with territory to roam, or a small one that requires little maintenance?
- ❏ **Spacious kitchen.** For many people, the kitchen is the heart of the home, and they simply couldn't be happy with a small kitchen.
- ❏ **Fireplace.** Many people, myself included, can't live without this.
- ❏ **Open staircase.** A touch of elegance that many people like.
- ❏ **Master bedroom suite.** Master bedrooms with private bathrooms are of strong appeal to most everyone.
- ❏ **Office or library.** Remember that any spare bedroom can serve this function.
- ❏ **Upstairs laundry.** Often overlooked, but greatly valued.
- ❏ **Spare bedroom.** Give careful thought to how many overnight guests you'll want to host.
- ❏ **Entry foyer and closet.** My wife lamented the lack of this feature for years.
- ❏ **Recreation room.** Especially desirable if you have kids or grandkids.
- ❏ **Great room.** Wide-open living spaces with vaulted ceilings are most often found in homes less than 20 years old.
- ❏ **Formal dining room.** Sometimes omitted in homes featuring great rooms.
- ❏ **Air exchanger.** If a family member suffers from severe allergies or asthma, this can be very important.
- ❏ **Fenced-in yard.** Very much in demand for families with small children or pets.
- ❏ **Deck or patio.** For families dedicated to outdoor living, this can be a crucial feature.
- ❏ **Pool.** Usually most appealing to people who have never had to maintain one. Think this through before deciding your new home must have one.

each day. In fact, setting aside a week of vacation for your house hunt is a great idea. In a fierce real estate market, you'll need to respond almost instantly if you find a home you truly like. This will be most practical if you've got plenty of free time for making telephone calls and meeting with agents, owners and bankers.

Statistics also tell us that a bid placed on a weekday has a greater chance of being accepted than one made on a weekend—probably because there is more competition on the weekends. This is another good reason to take some time off to do your house hunting and inspections during the week.

Begin the countdown. Gather the classified ads, flyers, and Internet printouts from homes in the neighborhoods you've selected. Next, highlight the homes that have the features you're looking for. Finally, get on the phone and begin scheduling visits. If homes are selling fast, you may not want to wait for a public open house showing.

Assemble your inspection kit (page 5), prepare the car, and get ready to roll.

If You're Selling:

Although there's not much you can do at the last moment to improve your entire neighborhood, you can do some research and be prepared to talk knowledgeably about the quality of the neighborhood, focusing on its good points. Have lots of information at hand, so you (or your real estate agent) can provide the kind of tidbits that might be of interest to buyers. A young couple, for example, might be interested in how many families with children are on the block, along with information about schools and parks nearby. An older buyer, on the other hand, might be more interested in public libraries and walking paths.

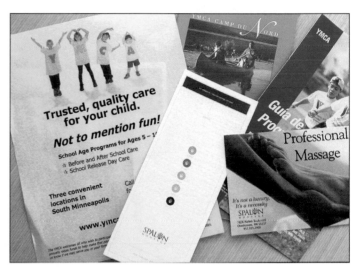

Keep a file on community programs and activities that might appeal to potential buyers. Your local library or community park building is a good source of this information. Your file can include information on things like:
- Community education classes
- Garden clubs
- Neighborhood picnic schedules
- Crime watch programs
- Home improvement funding programs
- Lawn and snow-removal services

Keep a file on any community activities and programs that might show prospective buyers that your house is in a desirable neighborhood.

LET THE SEARCH BEGIN: CHECKLIST

❏ Identify neighborhoods that suit your lifestyle needs.

❏ Explore selected neighborhoods, in person as well as by phone calls and on-line research.

❏ Determine what features are most important to you.

❏ Locate houses that have the right features and are in the right neighborhood.

❏ Schedule inspection visits (no more than two per day).

Chapter 3

View from the Curb

First Impressions

Now comes the fun part. Pulling up to the curb, you see the sign: *Open House, 12:00 to 4:00*. Or, if you've arranged for a private showing, the real estate agent is chauffeuring you to the house, or is standing on the lawn, grinning a welcome to you. But even now, there's no hurry. Anticipation is sweet. Take a moment and simply study the house from the street. Surprisingly often, you can see things without even opening the car door that will save you hours of time.

If you're doing a private showing with a real estate agent, don't feel that you need to hurry inside. Invite the agent to relax for a few minutes while you let your first impressions sink in.

Personally, I reject any home where I see a roof line that sags like the back of an old horse, a foundation that's clearly bulging or buckling, or stucco or brick siding with severe problems evident from the street. I look for large trees that are either too close to the house or that are species with known disease problems. Large elms are beautiful trees in the midwest, but should Dutch Elm disease ravage the tree—and there's a good chance of this—you may soon pay thousands of dollars to remove it. Finally, I don't like to see a steep sloped yard with a questionable retaining wall.

Notice whether the house has been altered through remodeling

Surprisingly often, you can see things without even opening the car door that will save you hours of time.

Save your time and inspect only those homes that genuinely appeal to you.

in a way that compromises its architectural style. Now, for many people (myself included) this isn't much of a problem. In fact, an idiosyncratic house can be very appealing, provided it has the quality-of-life features you need. But if you're a purist who looks at the home as an investment, a compromised design might reduce its ultimate resale value. This is especially true of homes in upscale or historic neighborhoods.

These types of problems are ones that can be very costly or even impossible to remedy. That's not to say that you shouldn't buy the house—just that you should be prepared for some very expensive

If you're a purist, you might steer away from a house with an addition that's ill-suited to the style of the house, such as an ungainly second-story expansion.

projects. Some people buy homes purely as fixer-uppers, with an eye toward resale at a substantial profit. Most of us, though, might consider these types of problems deal breakers.

It's also possible you'll spot something in the immediate proximity that tells you that there's no way you'd want to buy the home. If the house is adjacent to a gas station or liquor store, for example, it could be a waste of time to step out of the car. Personally, a home adjacent to a major highway or thoroughfare would scare me off since I'm not fond of noise and carbon monoxide. But I also see things that don't scare me at all, though they can frighten some people off.

Deal Killers:
If you have any health or safety concerns about the house, look elsewhere. Although the scientific jury is still out on this one, high-tension power lines make some people nervous.

The need for paint or minor cosmetic work is no big deal. Nor do I shy away from homes with landscaping that is a bit shabby. An unkempt lawn, shaggy shrubs—these things are pretty easy to fix. And if these simple cosmetic things scare off a certain percentage of prospective buyers, there will be less competition for you. Of course, it's entirely possible that these cosmetic aberations cover up more serious problems, but don't draw this conclusion automatically.

Deal Killers:
A house that has significant siding or roof damage will be facing some pretty extensive repairs, which means you may not want to waste valuable inspection time.

A large tree growing within a few feet of the house is likely to be troublesome in the future. It can cause damage to the roof and foundation, and will cost thousands of dollars to remove.

Never dismiss a house purely because your first impression isn't completely enthusiastic.

Before you go in, it's a good idea to drive away, around the block or through the alley, and simply look at the house and neighboring houses from all angles. A glance at the back yard from an alley can tell you volumes. What do you see when you glance at the back yard? Do you see signs that family pets have turned the lawn into, well, something closer to a barnyard? For that matter, have a glance at the condition of the immediate neighbors' back yards and utility areas. These will be the people you might call your own neighbors for quite a while. Is there play equipment that signals the presence of small children? This could be a good thing or a not-so-good thing, depending on your perspective.

Keep driving now, around to the front of the house. If possible, circle the block and approach the house from a different angle than you first did. Just observe the house as you pull up. Can you visualize this as your home in the condition you now see? Could you see it as your home with a few cosmetic improvements? Gut instincts are worth a lot here, so don't discount them without careful consideration.

Like what you see so far? Then it's time to walk up to the entry.

Signs of Gold:

A well-maintained house with an architectural style that fits the surrounding neighborhood is always a safe investment. Should you choose to sell in the future, such a home will maintain its value well.

This plantation-style colonial is ideally suited to its Atlanta neighborhood.

If You're Selling:

It's sometimes hard to see your own home the way strangers will see it. A good exercise is to take a Sunday afternoon to visit a few open houses other than your own. Keep your eyes wide open, and pay attention to the kinds of details that draw your eye. When you return home, continue the exercise with your own property. You'll be surprised by the kinds of details you spot.

When preparing for your open house, pay particular attention to the outside entryway. This will be a focal point for nearly everyone who visits your home. Investing a few dollars in gleaming new brass doorknob, knocker, and mailbox can be money very well spent. And for first impressions, nothing counts more than neatness. Edge the flower beds, clean weeds from sidewalk cracks, dust the light fixture on the porch, etc., etc. It can be well worth while to trim trees and shrubs, especially those that overhang sidewalks and steps.

Finally, a fresh watering of lawn and sidewalk creates shimmering highlights that make your front landscape very appealing.

The front entry draws the most attention from prospective buyers, so make sure it is neat and well tended.

VIEW FROM THE CURB: INSPECTION CHECKLIST

	Good	Average	Poor
House style (Fits neighborhood? Suits your taste?)	❏	❏	❏
Roof (Straight and smooth? Obvious damage or wear?)	❏	❏	❏
Siding (Appears weather tight?)	❏	❏	❏
Foundation (Appears sturdy? Sagging or buckling?)	❏	❏	❏
Sidewalk, steps, & concrete (Good condition? Buckling?)	❏	❏	❏
Trees (Well trimmed? Too close to house?)	❏	❏	❏
Landscape (Well tended? Needs major work?)	❏	❏	❏

Chapter 4

Meeting Your Host
Starting Off on the Right Foot

Start off on the right foot by establishing an understanding with the real estate agent or owner of the house.

Your inspection will be much smoother, and the agent or owner will be much more agreeable and helpful, if you start off on the right foot.

Provided the curb appeal was enough to make you want to see more, then the next order of business is meeting the real estate agent or the owner who is representing the home. If your first visit to the house is a private showing, your meeting will be one-on-one with the agent or owner. At public open-house showings, things are a bit less personal since there probably will be more people milling about. Either way, your inspection will be much smoother, and the agent or owner will be much more agreeable and helpful, if you start off on the right foot.

Here's the typical scenario:

As you enter, the greeter will welcome you and probably offer you a real estate fact sheet, and perhaps a treat or beverage. Take the fact sheet; politely turn down the treats until after you've completed your inspection. You'll need your hands free as you explore the house. If you're visiting a public showing, you'll be asked to sign a guest register, and you should expect the agent or owner to follow up with you a day or so later. He or she will ask you something about where you now live, and what your housing needs are—you're being scoped out to see if you are a serious prospect.

If you happen to be visiting for recreational interest and have no serious interest, as I often do, explain this to the agent or owner. It will save him or her the wasted time of calling you later. Most hosts will have no trouble letting you browse in a casual way.

The Paper Trail

In some cases, you'll not see much more than a simple real estate fact sheet when you begin your inspection, but in other cases two or three other important documents may also be available for you to look at.

The **real estate fact sheet** is a simple document that you're almost certain to receive. It can take a variety of forms, from a simple black-and-white typed sheet of information, to a multipage brochure complete with a portfolio of color photos.

In addition to this fact sheet, ask if there is a ***truth-in-housing***

HomeAvenue.com   Featured Listing

Price: $485,000
1000 High Street
Minneapolis, Minnesota 55417
Contact Owner: John Smith
Phone: 612-555-2345
Email: jsmith@aol.com
HomeAvenue ID #: 7365

Bedrooms: 4
Bathrooms: 2
Finished Sq. Ft: 3300

Spacious 4 bedroom executive home. Large corner lot. High Street location. New tile roof. Immaculate!

Property Data:

Year Built:	1929	Garage:	Two Car
Style:	Two Stories	Heat:	Hot Water
Exterior:	Stucco	Air Conditioning:	Window
Lot Size:	60 x 140	Fuel:	Natural Gas
Foundation Size:	1300	Fireplace(s):	Two
Basement:	Full	Taxes:	$3,500

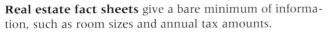

Real estate fact sheets give a bare minimum of information, such as room sizes and annual tax amounts.

EXTERIOR DESCRIPTION		FOUNDATION	
Foundation	Conc. Blk	Slab	
Exterior Walls	Wood	Crawl Space	
Roof Surface	Asphalt	Basement	
Gutters & Dwnspts.	Yes	Sump Pump	N
Window Type	D-Hung	Dampness	Not
Storm/Screens	Yes	Settlement	Not
Manufactured House	No	Infestation	Not

Dining	Kitchen	Den	Family Rm.	Rec. Rm.
		2	1	
1	1			

The appraisal report gives lots of details, but makes no value judgements about the condition of the various features of the house.

statement available. In some states, this report is required by law. Generally, the truth-in-housing statement is the result of a fairly cursory professional inspection, and may give you a few hints about problems the house may have. But don't expect this report to take the place of your own in-depth inspection, or an inspection by an accredited inspector.

In addition, it's possible that the agent or homeowner has an *appraisal report* that was used to determine the appropriate market value for the home before they put it up for sale. The appraisal is made by comparing the floor area of the home and its features with other comparable homes in the general geographic area. Lenders often require an appraisal inspection before they will agree to lend money to purchase a home. If you're borrowing money to buy a house, you may well end up paying for this report yourself, but it's not uncommon for sellers to have such an appraisal already done. You are most likely to see this report if the appraised value supports the asking price of the home, but in any case, ask if an appraisal has been done. An appraisal report is generally much briefer than a full-blown professional inspection report. When an appraisal report has already been done, it's likely the agent or owner will have a single copy that you can glance at, but not enough copies for you to take one home. If you're serious about the house, ask for a photocopy of the appraisal report. It's full of information that will be helpful as you later evaluate the home and negotiate a purchase price.

Signs of Gold:

A professional inspector's report will list any deficiencies that the inspector noted. A homeowner who provides a professional inspection report clearly has nothing to hide.

In rare instances, you may run into a seller who has had a professional inspection done on his or her house. This is the same kind of inspection you might want to purchase for yourself at a cost of $200 to $400, so pay attention if a copy of this report is available for you to look at. Make sure, though, that the report is up to date—not the same report the homeowner purchased when he or she bought the home eight or ten years earlier. Here, too, it's unlikely you'll be able to have your own copy of the inspection report, so study it carefully and take notes. I've been known to snap a high-resolution digital photograph of key pages in an inspection report, which I can later print out for my own benefit.

Beginning Your Inspection

Before setting off to explore the home, explain your goals to the agent or homeowner. Because you've arrived with your inspection kit of tools and testers, notepads and camera, you won't fit into the traditional mold of prospective buyer. Very few home buyers are this well prepared when they arrive at a showing, so your seriousness will sometimes catch the owner or agent a bit off guard. Not to worry; just explain that you want to look very closely at the house, and explain what you'll be doing.

Make sure you ask before taking photos. Taking snapshots in particular seems to worry some people, so if you run into opposition, respect these wishes and take written notes instead. It's also possible that once you establish a relationship with the agent or owner, he or she will become more willing to let you take a few photos.

Especially at a private showing, the agent or owner will be eager to be helpful, but it really is best if you establish a little bit of space so you can view things quietly and at your leisure. Spend a few minutes in preliminary conversation, then politely but firmly state that you'd like to look around on your own. Ask specifically for permission to peek into closets, crawl spaces, the garage, and the attic. Sometimes these areas are considered off limits at a public open-house showing. If so, you'll want to arrange for a private follow-up visit, because you definitely will want to see all these things at some point.

Especially if there aren't many guests, the host may want to walk along with you like a long-lost friend. I rather dislike this, and so try to insist on a bit of independence as I make my inspection. Homeowners showing their own homes are especially "friendly" in this way.

Open houses hosted by owners are a mixed blessing. Owners know many details about the home that a real estate agent won't,

Explain to the host what you'll be doing with the inspection tools you've brought along, such as this circuit tester.

Ask specifically for permission to peek into closets, crawl spaces, the garage, and the attic.

but they can also be quite nervous and much too eager to help you see the merits of the home. Just be firm in a good-natured way about your need to view the house without distraction.

Sometimes, giving the owners 10 or 15 minutes to tell you the house's story is enough to relax them into leaving you alone. Later, there will come a time when you want to talk with the owner in great detail—but the initial open house is not the place for it. Assure the greeter that you want to talk to them at some length at the end of your visit. Be quite firm in this regard. There are a few real estate agents who take an approach like a used-car salesman. It's very difficult to inspect things adequately with someone hanging over your shoulder.

With these formalities out of the way, you're ready to begin the official inspection. Everyone has a different method for approaching a house. I like to start with the exterior and move my way toward the inner world, so my habit at this point is to tell the owner or agent that I'm going to begin by looking over the yard and exterior of the home.

Sometimes, giving the owners 10 or 15 minutes to tell you the house's story is enough to relax them.

MEETING YOUR HOST: INSPECTION CHECKLIST

❏ Take a copy of the real estate fact sheet.

❏ Is there a truth in housing statement available for you to look at?

❏ Is there an appraisal report you can review?

❏ Has there been a professional inspection done? If not, is the owner willing to allow you hire an inspector?

OPEN TO AGENTS
DORSEY
ALSTON
(¼)555-0000

Inspecting Townhomes, Condominiums, and Planned Unit Developments

If the home you're inspecting is part of a planned development, or a townhome or condominium complex, you'll inspect things in much the same way as with a traditional single-family home. There are, however, a few special considerations having to do with the nature of such communities.

Condominiums. A condominium is essentially an apartment that you purchase, where you own all elements that extend inward from the walls, ceiling, and floors. The exterior walls, roof, and outbuildings are owned in conjunction with all the other owners in the complex. The land, however, doesn't belong to you in any way. Beware of condo complexes where only a small percentage of the units are sold and the rest rented or leased.

Townhomes. A townhome can be defined as a single-family home that shares common walls with one or more other homes, where you also own the land on which the home sits—which is the major difference between a condo and a townhome. Thus, a townhome is generally less expensive than a traditional single-family home, but more expensive than a condominium.

Real estate professionals say that it's best to buy townhomes that are in the mid-range of the pricing structure for the development. These units tend to be the best investment. Also, it's generally best to avoid units that have been remodeled to be very different in appearance than others in the complex, because they can be hard to sell in the future.

Planned Unit Developments. In PUD communities, the homes are traditional single-family homes, but they are sold as part of a homeowners' cooperative, and thus are subject to maintenance fees, association fees, and certain restrictions.

The main advantages of PUDs are that they try to carefully control the behavior of residents, and that they often feature attractive recreational facilities—golf courses, swimming pools, community centers. You do, however, pay for these amenities. And you'll be responsible for all the normal maintenance and repairs to your home and landscape—work that can be forced on you by the homeowners' association.

In all three types of planned housing scenarios, it's likely you'll instantly become part of an association with the other homeowners. The most notable features of these organizations are the covenants, conditions and restrictions (CC & Rs) that dictate how the property can be used. Many people are reassured by living in a community where no one can be too eccentric, while others find that they chafe badly under the burden of such restrictions. It's essential that you make sure that the restrictions work for you. Some of the issues often dictated by CC & Rs:

- Paint colors for home exteriors
- Color of curtains and blinds visible from the street
- Parking of trailers or RVs in driveways
- Parking of cars in driveway
- Species of trees, flowers and groundcovers in landscaping
- Fencing materials and heights
- Swings and playsets
- Outdoor clotheslines
- Pets
- Style of mailbox
- Wreaths and holiday decorations
- Ambient noise

Owners in a condo, townhome, or planned unit development pay a monthly association fee or become part of the homeowners' association, which meets regularly to determine policy for the community. It's a great idea to attend one of the monthly meetings if you're thinking about buying into such a community. If this isn't possible, have a look at the association's bylaws and the transcripts of some of its meetings.

As part of the homeowners' association, you'll be paying a monthly fee for maintenance of the shared property and the running of the organization. You'll be paying quite a bit of money, so make sure you're getting your money's worth.

A townhome is basically a single-family home that shares common walls with one or more other homes.

Questions to ask:

In condominum communities
- How many units are owned, and how many rented in the complex? Ownership of 60% or more is best.
- Are there any planned special assessments? Major landscaping work, new roofing, and other major repairs are typically passed along to the individual owners in condo developments.

In townhome communities
- Is the homeowners' association involved in any litigation?
- How does the unit you're inspecting compare in price to others in the development?
- How much of the exterior maintenance work is your responsibility, and how much is covered by the association fees?

In planned unit developments
- Can the homeowners' association force you to make external repairs and improvements?
- Is the developer of the community financially sound? Ask about other developments in progress.

Regarding the CC & Rs
- What is the procedure for obtaining variances from the established covenants?

Regarding the homeowners' association
- How have the assoication fees changed over the last few years? Beware of increases in excess of 20%.
- Does the association have a surplus fund to cover emergency repairs?
- Have the elected leaders served for long stretches (a good sign)? Or has there been frequent turnover (a bad sign)?

If You're Selling:

To Decorate or Not to Decorate

Some sellers invest considerable money and time in decorating for an open-house showing. I've even known folks to invest in new furniture in order to impress prospective buyers. Unless your home has been decorated in a way that might be unpleasant to people, it's usually a waste of time and energy to redecorate in a major way. The one exception is for a very upscale, expensive home. At the higher end of the real estate market, it might even make sense to hire an interior decorator to prep your home for showing.

Generally, the less furniture and accessories present in the home at the time of your open-house showing, the better. It's even a good idea to put as much furniture as possible into storage. The more space visitors feel, the easier it will be for them to visualize the house as they want it. In fact, some real estate agents say that a completely empty house shows better than one filled with furnishings. At the very least, it may make sense to get rid of badly worn furniture and excessive accessories. Many real estate agents advise sellers to have their estate sales and garage sales well before the open house, simply to get excess out of the way.

If you do redecorate, use neutral colors—whites and off-whites. Lighter colors simply feel larger and more spacious, and will offend no one's sensibilities.

Remove all signs of family pets. Even animal lovers can be put off by the smell of a litter box.

Create a focal point. Even though most advice says to make your home look generic, I believe it's very helpful to have one element in your home that is quite unusual and that will be remembered in a positive way. This is just good marketing—creating a point of difference that makes your home stand apart from others. It might be a distinctive painting, or a small display for distinctive mementos.

Photo Courtesy of Kolbe & Kolbe Millwork Co., Inc.

Keep furnishings and decorations to a minimum, and maximize light by keeping draperies open, or omitting them altogether.

Being a Gracious Host

There are, of course, many, many tricks for making a house seem welcoming to guests—a fire in the fireplace, freshly baked bread on the counter, fresh cookies and coffee. All these things are somewhat effective, but will also be quite obvious to the prepared visitor. An often-overlooked strategy is to do whatever it takes to increase the amount of light in your home—installing light bulbs of higher wattages, for example. Light colors, open drapes, and mirrors on the walls all make your home brighter and more appealing.

It can be a bit hard to bite your tongue and just allow people to roam about—but this is what is most effective.

If you will be hosting the visitors yourself, the best strategy is to simply be friendly and welcoming, and to give your guests space to see the house as they see fit. It can be a bit hard to bite your tongue and just allow people to roam about—but this is what is most effective. If people do seek answers to questions, then be as forthcoming as possible. But wait for the visitors to come to you. Avoid the aggressive salesman mode—pressing at every opportunity to force a sale.

Give yourself peace of mind by putting all small items of value safely in storage on the day of your showing. Although they're rare, visitors with sticky fingers do exist.

CHECKLIST FOR HOSTS

❏ Remove or store excess furniture and accessories, and any that are severely worn or damaged.

❏ Make copies of any appraisals or inspections that have been done.

❏ Create a file or bulletin board with flyers, neighborhood park programs, and community education programs and classes.

❏ Create a small scrapbook or bulletin board with photos of the house at different seasons of the year.

❏ Find temporary lodging for all pets, no matter how soft and cuddly—and remove all litter boxes and other pet accessories.

❏ Clean, clean, clean.

❏ Pay close attention to cleaning shiny surfaces—gleaming highlights put your house in the best possible light.

❏ Organize closets and utility areas. Savvy shoppers will be looking at these spaces, too.

❏ Maximize light by installing high-wattage bulbs and keeping drapes wide open.

❏ Secure jewelry, collections, and any other small valuables.

Chapter 5

Welcome to the Great Outdoors
Assessing the Yard & Landscape

This beautiful front yard landscape includes a drip-irrigation system, lawn edgings, and mulched planting beds, making it very easy to care for.

The landscape can contribute 15% to 20% to the home's value, and its effect on quality of life is even more important.

I like to start an inspection with the outside of the house—the landscape—although some people prefer to make this the last stop on the tour. Either way, be leisurely about this part of the inspection, because it's quite important. Real estate professionals say that the landscape can contribute 15% to 20% to the home's value, and its effect on quality of life is even more important. Over the last 20 years, the yard has become an essential part of our recreational and social lives.

For landscape professionals, a landscape consists of both "softscape" elements—the growing plant materials—and "hardscape" elements: fences, patios, walkways, arbors, decks, and other nonliving components. It's a convenient way to look at the landscape for inspection purposes, as well.

Begin with the Softscape

Walk back out to the street or the edge of the property and look at the total effect of the living landscape. Your lifestyle will play a big part in how you view the yard and landscape. If you're an outdoor person who sees yard and garden work as pleasant exercise and who enjoys outdoor activities, then a large yard with stately shade trees, big garden beds, a spacious deck and patio, and tall privacy fences will be strong selling points. But if you're an indoor person, you may not want to pay for and maintain these things. So the first question to ask yourself is: does this yard meet my needs and my lifestyle?

Evaluate the grade of the site. A yard need not be completely flat and level, but a lawn that's very uneven will be hellish to care for. Most important, look for evidence that the ground slopes away from the foundation of the house. In all but the driest climates, a grade that directs rainfall toward a foundation is problematic. Within 10 feet or so of the foundation, the ground should slope away from the foundation about ¾ inches for each horizontal foot.

It's a good sign if the yard incorporates a shallow ditch-like depression in low-lying areas to channel water away from the foundation. As you walk around the yard, you may see drain grids or an outflow pipe, indications that a dry well system has been installed to carry water safely away from the house.

Signs of Gold:

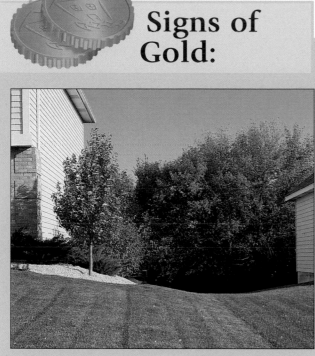

A shallow ditch like this one, called a swale (left), will channel water away from the house rather than toward the foundation. You may also find signs of a below-ground drainage system, such as this outflow pipe (right). Both are signs of a landscape that was well designed, probably by a landscape architect.

Inspect the trees. Stately shade trees are a great feature if they're in good condition and well cared for, but they can be a decided liability of they're not healthy and well placed. The huge ash tree shown at right casts a lot of shade over the house below it, but it also has a suspicious crack running up to the crotch. The tree could cost two or three thousand dollars to remove, and could cause tens of thousands of dollars in damage should it be blown over in a wind storm. Any major tree that lies within 10 to 15 feet of the house foundation is a potential source of trouble. Though it's rare, tree roots can exert enough pressure to actually crack and buckle a foundation or break sewer lines. Swaying or broken branches can damage a roof.

Understand that trees have a life expectancy just like any species, and certain trees are not very long-lived. Poplars, aspens, cottonwoods, soft maples, and many pines are among short-lived tree species. They're often planted because they grow quickly and provide quick shade, but beware if the yard has large specimens of these tree species; you may be looking at major removal costs in a few years. In addition, these trees can be somewhat soft and brittle, which makes them susceptible to wind damage.

This tree has a suspicious crack— and because the tree is very close to the home, It holds the potential for a serious problem.

Diseased elm trees are often marked with spray paint by the city forestry service to tag them for removal. If you see tagged elms in the neighborhood and the home you're inspecting has elm trees, they will be at severe risk in the near future.

Better landscape treees are those that are slower growing, have relatively hard woods, and are indigenous to the region you're in. In the midwest, for example, a yard landscaped with oaks, hard maples, and lindens will be less problematic than one landscaped with magnolias and fruit trees.

Although everybody likes the look of a tree with thick, dense foliage, in reality it's a better sign if the tree has been routinely thinned so that there's plenty of air moving through its branches. A well-trimmed tree is nearly always healthier, and it is much less likely to suffer wind damage during storms. In fact, arborists say that a tree should be pruned to remove fully a third of its branches every few years. On a large tree, look for the grown-over scars that indicate the tree has been routinely pruned.

Finally, you need to pay particular attention to the species of tree. In many parts of the country, major disease or pest problems plague certain tree species, and if the yard has several trees of such a species, you could be in for many thousands of dollars in community-mandated removal costs. Some common tree diseases are listed in the chart below, but if the yard has many trees, it makes perfect sense to have an arborist inspect them and give them a clean bill of health.

Although everybody likes the look of a tree with thick, dense foliage, in reality it's a better sign if the tree has been routinely thinned.

TREES PRONE TO CATASTROPHIC DISEASE OR PEST INFESTATIONS		
Tree	**Disease or Problem**	**Region Affected**
Elm	Dutch elm disease	North America
Oak	Oak wilt	Eastern U.S. and Great Lakes region
Dogwood	Dogwood anthracnose	Eastern North America
Palm	Ganoderma butt rot	Florida
Pine	Southern pine beetle	Southern U.S.
	Sawfly	Southeast Canada, northeast and north central U.S.

Elm trees are prone to Dutch elm disease across virtually all of North America. If the property you're considering has elm trees, look around the neighborhood for signs that other elms are sickly or tagged for removal. This is a strong sign that Dutch elm has hit the neighborhood, and that your own trees will be at risk.

Here's a good tip: any tree that has toadstools or other fungal growths sprouting out of the ground at its base is at great risk. The fungus is feeding on dead wood material in the roots or base of the tree, indicating that the tree is dying. If you're inspecting in the autumn, beware of any tree that seems to be changing leaf color earlier than other trees in the neighborhood. Early color change is a sign of stress; this tree may be in the early phases of a killing disease or condition.

It's possible to preventatively treat a tree to improve its resistance to a disease like Dutch elm, but it's a good idea to see paperwork to verify how the trees have been treated. The jury is still out on the actual effectiveness of these treatments.

Finally, with trees of moderately young age, look at the shape of the lower trunk of the tree. It should be slightly flared as it goes into the ground. It is quite common for homeowners or even professional nurserymen to plant trees too deep. It can take 5 to 7 years for such a tree to smother, but it will eventually die, even if it looks healthy at the moment. If the trunk has a good flare, with no indication of roots girdling its base, this is a good sign.

Inspect the lawn or groundcover. Turf grass is the groundcover of choice in most parts of the country, but it's possible you'll be looking at a ground ivy or another living ground cover when you inspect the house. Whatever the groundcover, look at it closely. The owners have probably mowed and weeded to put the yard in its best possible light, but if you get down on bended knee, you'll be able to see the relative health of the plant life. A good lawn or groundcover will appear uniform in texture, while a bad lawn will show evidence of weeds and crab grasses.

Pay attention to the lawn's size. For some people (I'm one of them) mowing and tending plantings is exercise, recreation, and therapy. If this describes you, a big lawn is a plus. For others, lawn care is sheer drudgery and a large lawn is a decided negative; if so, you'll be more drawn to a yard with large areas covered with patios, decks, or mulched garden beds.

If the owner was ambitious, it's possible he or she resodded problem areas of the yard—usually shady spots. Although this makes the yard show well to the casual buyer, sod installed in shady areas will never live very long. A better sign is a yard where

Just because it's green doesn't mean it's pastoral. Have a close look at the quality of the lawn. Is the green you see the result of dense turf grass (top), or is there a fair percentage of creeping charlie, chickweed or other undesirables mixed in (bottom)?

Signs of Gold:

A shady yard planted with shade-loving plants and groundcovers will be much easier to care for than one where turf grass has been bullied into terrain that doesn't suit it.

An irrigation system will save much yard-care time, especially in arid climates.

the shady areas have been overlaid with garden beds planted with shade tolerant shrubs and flowers.

In arid climates, it is much easier to care for groundcovers that don't require a lot of watering. Here, look for ground-covers featuring indigenous plants rather than turf grass, and the presence of a carefully controlled irrigation system. Without this, you'll probably find yourself watering manually on a daily basis. Keeping a traditional turf lawn in good health is a very difficult and expensive challenge in these climates, and is nearly impossible without a watering system.

Look at the shrubs and flowers. These are the ornaments of a living landscape, but don't downplay the importance of decorative plantings. Shrubs can offer a wind and noise buffer, as well as improving privacy, while flowers make a yard much more appealing.

Inspect shrubs to see if the foliage is full and healthy. Sparse leaves or yellowing foliage are bad signs. Shrubs are expensive to replace, so they deserve some scrutiny. Like trees, shrubs need to be pruned regularly, so look closely for pruning scars indicating they've been well cared for.

Flower beds planted with perennials, such as daylily, iris, and daisy species, will not need to be replanted yearly, while beds filled with annual flowers will need to be replanted every year. Any planting area covered with landscape fabric and a thick bed of shredded wood or other mulch will be easier to care for. And the best sign of all will be a timed drip-irrigation system that puts carefully controlled amounts of water right where they're needed.

Inspect the Hardscape

Now it's time to inspect the structural elements of the yard, those elements commonly known as the hardscape, which are built from wood, masonry, and other materials.

Does the yard have retaining walls? The timber, concrete, or stone retaining walls that create terraces in sloped yards are one of the most crucial aspects of any landscape. A large retaining wall with serious buckling can cost thousands—or tens of thousands—of dollars to replace. And replacement may be ordered by city authorities, if the structure is deemed to be a safety risk. The contractor bid to repair the extensive retaining wall shown here was $50,000. Unless this issue was factored into the asking price, such a problem could be deal breaker.

In general, timber retaining walls have the shortest life span, while an interlocking block wall, if properly installed, can last for 25 years or more.

Deal Killers: A large retaining wall with serious buckling is one of the most expensive of all home repair projects—and it's not one you can avoid forever.

RETAINING WALL LIFE EXPECTANCY & REPLACEMENT COST		
Material Type	**Life Expectancy**	**Replacement Cost (installed)**
Timber	10 years	$10 per sq. ft.
Concrete	20 years	$180 per lin. ft.
Interlocking block	25 + years	$13 per sq. ft.

Inspect walkways and other masonry structures. Some small cracks and unevenness in poured concrete slabs, brick surfaces, and steps are normal, but be on the lookout for signs of substantial heaving or settling. Look closely at any walkways or concrete slabs abutting the house or garage. Because the soil used to backfill around a foundation is often less dense than surrounding soil, it's pretty common for it to settle. A buckled, slanted sidewalk or driveway can channel water into the home. The sinking shown here will channel rain water into the garage. If you bid on a house with these problems, keep the cost of repairs in mind. Although these problems can be lived with for several years, correcting them can be very expensive if you hire a contractor—or

Tree roots causing a city sidewalk to heave can lead to tax assessements to pay for repairs.

difficult and time-consuming to do yourself.

The city sidewalk can also pose problems for you. You could be liable for repairs to any sidewalk that has buckled due to pressure from tree roots, for example. If the city orders repairs, you likely will be billed $200 to $300 for each panel of sidewalk that needs to be replaced.

- Concrete slab replacement cost: $3.50 or more per sq. ft.

- Concrete step repacement cost: $35 per lin. ft.

A badly buckled, sloped driveway or patio slab can allow water to flow down along the foundation. These problems are difficult and expensive to fix.

With brick-paved patios, inspect the surface to make sure it is flat and isn't buckling. If the joints are mortared, check to make sure the mortar is secure and in good shape.

Tracks of the Poor Craftsman:

This irregularly repaired concrete stairway is clearly substandard. The vertical rise varies radically from step to step, and will probably be flagged as hazardous by any building inspector who sees it.

Water flowing over a sidewalk slab from this adjoining hillside runs directly into the house foundation. The sagging sidewalk is a good indication that a problem exists.

Signs of Gold:

A newer retaining wall built with interlocking concrete block is a good sign. This wall has seepage holes to allow for water drainage. Such a wall is unlikely to have any problems whatsoever for many, many years.

A newer deck with posts of pressure-treated lumber resting on concrete foundations should provide good support for many years.

Inspect fences. Most fences are made of wood, and so are subject to decay, especially at the points where the posts meet the ground. Ideally, the fence should be made of pressure-treated lumber or cedar to make it weather resistant, but this can be hard to determine, especially if the fence is painted. Visually inspect the bottoms of posts, and give the fence a firm shake at various points to see if it is solid. A wood fence overgrown with vines might look very nice, but moisture trapped by the foliage can wreak havoc on wood, as shown in the photo at right.

Newer fences may be made of vinyl. If it's sturdy, such a fence will last a very long time and require almost no maintenance. A well-installed vinyl fence is a sign of gold.

• Wood fence replacement costs: $12.00 or more per lin. ft.

Vines and wood fences don't mix. Always look beneath vines to check the condition of the posts, stringers and siding. Wood that is whitish or grayish in color is rotting.

Frost heave can cause major damage to a deck. Here, footings that were too shallow caused the center row of posts and beam to heave enough to render the deck unusable. The outer posts have actually lifted away from the footings. Most signs of frost heave will be more subtle than these.

A decayed decking board isn't hard or expensive to replace, but it may mask decay to posts and joists that support the deck, which is a much more serious matter.

Finally, look at the deck. Check both the surface boards and the underlying structural members. Indications of decay or rot will be pretty evident; probe the wood with an awl to check for softness. Except in the driest of climates, all decks need to be treated with a waterproofing agent, and failure to do this will be quite obvious.

In areas with freezing winters, the foundations for deck posts must be sunk below the depth where frost reaches. If the footings are improperly dug, freezing water will heave the footings and can skew the deck off level. In the extreme example shown above, the heaving of the center post footings has lifted the outer posts entirely off their footings and placed the deck at an angle that leaves it unusable. This deck will cost many thousands of dollars to replace. It's unlikely you'll see a situation quite this bad, but if the home you're considering has a deck, check the post footings carefully to see if there is any sign the posts have separated from the footings, or if the footing themselves have begun to tilt.

Check posts for signs of rot. A few areas of decay aren't a big deal, but a deck or fence with many rotted posts will require attention in a year or two—if not sooner. Deck posts should be resting on concrete piers, not buried directly in the soil. Direct contact with soil will eventually rot any wood. In a wooden fence, replacing a post isn't too hard—but replacing deck posts and joists is another matter.

• Deck construction cost: $23 or more per sq. ft.

If You're Selling:

A few hundred dollars spent in adding healthy flowers and shrubs can dramatically increase the open-house appeal of your home. If your lawn has bare spots, invest in some new sod to cover these areas. Cutting a neat edge along lawn and flowerbed boundaries—or installing plastic edging—will make your landscape look very well kept, and will probably coax more people into your home to have a closer look.

If you haven't the time to invest in lawn improvements, then make sure the lawn is at least trimmed quite short on the day of the open-house showing. It will appear at its best if the grasses and other plants are clipped close to the ground.

If you haven't the time or budget for serious landscaping, then invest in some container plantings to dress up the deck, patio, and front entry. Your house will show in its best light, and the containers can be taken with you when you move.

If you haven't the time to invest in lawn improvements, then make sure the lawn is at least trimmed quite short on the day of the open-house showing.

Place growing flowers in containers on the deck or patio and around entryways to improve the appeal of your landscape.

Welcome to the Great Outdoors: Landscape Checklist

Item	Good indications	Condition		
		Good	Average	Poor
Yard slope	• Flat, level	☐	☐	☐
	• Slopes away from foundation	☐	☐	☐
Trees	• Good species for site	☐	☐	☐
	• Well trimmed, healthy	☐	☐	☐
	• Well positioned on lot	☐	☐	☐
Lawn or groundcover	• Healthy, few weeds	☐	☐	☐
	• Appropriate to sun exposure	☐	☐	☐
Shrubs & flowers	• Mulched and edged planting beds	☐	☐	☐
	• Perennials, not annuals	☐	☐	☐
Retaining walls	• Interlocking block rather than timbers	☐	☐	☐
	• No buckling or sagging	☐	☐	☐
	• Drainage features	☐	☐	☐
Walkways & patios	• No settling	☐	☐	☐
	• No buckling, heaving	☐	☐	☐
Fences	• Vinyl rather than wood	☐	☐	☐
	• If wood, no signs of rot	☐	☐	☐
Deck	• Posts elevated on concrete piers	☐	☐	☐
	• No signs of decay	☐	☐	☐
	• Sturdy railings & steps	☐	☐	☐

Anatomy of a Landscape

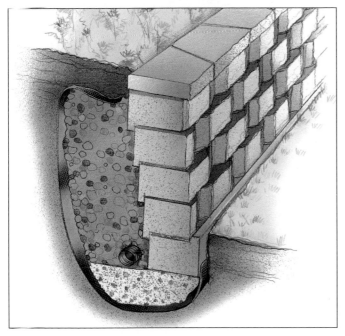

A well-installed retaining wall will include the features shown here. Most crucial is the presence of a sturdy gravel base, which you may be able to identify by probing around the base of the wall.

A brick patio built on a well-packed bed of sand will be very durable.

Wood fences should have posts with ample concrete footings to hold them securely in place. Metal connectors make it easy to replace damaged or decayed framing members.

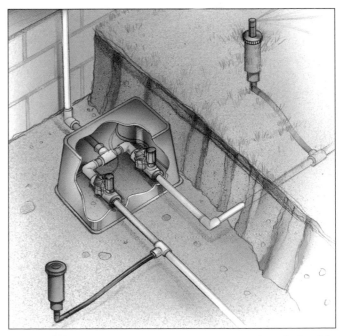

A sprinkler system should be easy to control, and should cover all parts of the yard and landscape.

Chapter 6

Regarding Roofs
Evaluating the Roof System

Signs of Gold: Cedar shingling is a good roofing material, and if you see a roof with a distinctive reddish-brown color, it means you're looking at a very new roof.

As you finish with your first inspection of the landscape and turn your attention to the house, begin at the top—with the roof. Take note of the type of roof you see, and make sure to later ask the agent or owner when the roof was installed.

With any house you're considering, severe water damage is one of the few problems that might instantly convince you to look elsewhere. The roof is where many such problems start, so pay close attention to this system. Water that penetrates the outer skin of a home can cause the hidden sheathing and framing to decay, and if not addressed, water damage can literally destroy a home.

When inspecting roofs, I find a small pair of binoculars is very helpful, since very few homeowners or real estate agents will allow you to get onto the roof surface to inspect it.

Look first along the spine of the house—the very top of the ridge—and look to see if there is any sign of sagging. The spine should be perfectly straight. A slight sag isn't a serious problem, especially on an older home, but if the spine sags badly, you may want to carry this inspection no further. Provided you don't see this kind of problem, now study the roof surface itself. There are several types of roof you'll commonly see.

With any house you're considering, one of the few problems that might instantly convince you to look elsewhere is the presence of severe water damage.

A COMPARISON OF COMMON ROOFING MATERIALS		
Roofing material	**Life Expectancy**	**Replacement Cost**
Granular shingles	20-25 years	$.80 to $1.50 per sq. ft.
Wood shingles, shakes	25-30 years	$2.80 to $3.20 per sq. ft.
Metal raised seam	60 years or more	$2.00 per sq. ft.
Concrete or clay tiles	75 years or more	$6.50 to $11.00 per sq. ft.
Slate tile	75 years or more	$8.00 to $14.00 per sq. ft.

Granular Shingles

The most common roofing material is granular shingles, identified by the hard mineral coating on the surface of asphalt- or fiberglass-based shingles. A granular shingle roof is pretty easy to evaluate, because there are some obvious signs it is nearing the end of its life.

Look for evidence of curling at the edges of the shingles. This always indicates aging. Also look for areas where the granular surface has worn away, leaving the darker asphalt core showing through. Finally, look in the gutters, or on the ground at the end of the downspouts. If you see a lot of granular shingle material, it's another bad sign.

Homes more than 20 years old may well have been reroofed. If possible, try to find out if the new roof was laid over the top of the old shingles, or whether a "rip" was done to remove the old shingles before reshingling. The best way to judge this is by closely inspecting the edge of the roof, where multiple layers will be very obvious. Most building codes allow two layers of granualar shingles, but not three. If the home you're considering already has two layers, this means that the next time you shingle, you will need to have the previous layers removed at extra expense.

It's rare to find a roof this bad, but anytime you see lower shingles that have worn bare of their mineral layer, the roof is on its last legs.

Signs of Gold:

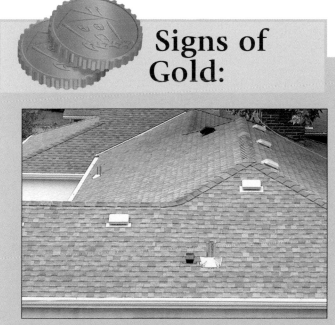

Shiny metal flashings and vents indicate that the house was recently reroofed. This roof will give decades of good service.

Closely examine the edges of the roof, and look for the presence of a metal or vinyl "drip edge." Drip edges protect the fascia boards from water damage.

Deal Killers: Some builders seeking to save money will lay a decking material of thin, ¼" plywood rather than the recommended ½" plywood. If this is the case, you may very well see visible sagging between roof rafters on a roof that appears relatively new, as shown in this photo. This is a pretty serious problem because the only fix is to completely remove the roof and install all new decking and shingles.

Because the roof system is so crucial, be extra cautious if you suspect it was installed incorrectly.

In cold climates, building codes now require that roofs be installed with a layer of impermeable rubberized fabric, usually called "ice-guard." In most cases, though, it's impossible to see this layer in a visual inspection, because building paper under the shingles obscures the ice-guard layer. It's wise to ask the owner if the roof includes an ice-guard membrane, and to inspect any documentation from the last roofing job. If the roofing job was substandard, it's possible that the code-required membrane wasn't installed. Here are some signs of a shoddy roof job:

- Uneven edges on the rows of shingles.
- Ragged trim edges along flashings and drip edges (See Tracks of the Poor Craftsman, page 46).
- New shingles laid over two existing layers.

Because the roof system is so crucial, be extra cautious if you suspect it was installed incorrectly.

Wood Roofs

Wood roofs come in two varieties—*shingles*, which are sawed by machine, and *shakes*, which are hand-split and have a rough texture. Most are made from cedar or redwood, though some are made from southern pine. Good quality wood shingles and shakes are durable, but the quality of the wood varies widely. Lower-grade roofs can begin to deteriorate within 5 or 10 years. Look very closely at a wood roof for signs of cracking and for curling. Although wood-shingle roofs can be very attractive, be aware that they are flammable, which can be an important consideration where wildfires are common.

Metal Roofs

These are usually called "standing seam" roofs, because the seams between metal panels are raised in interlocking seams. Copper, tin, and stainless steel are all used for standing seam roofs. These roofs were very popular 80 years ago or so, and are seeing renewed popularity, especially in rural areas with heavy snowfall, or where brush or forest fires are a signficant danger. Because there are no exposed nails in a standing seam roof, it's very durable. Don't worry too much about simple discoloration—the metals don't rust, and have been known to last more than 75 years, provided the surface isn't punctured.

Clay or Cement Tile Shingles

These types of roofs are easy to identify, because the tiles have rounded surfaces and are usually reddish in color. Older tile roofs are made of clay or terra-cotta. Newer tile roofs are usually fired clay or concrete. The color of concrete roofs tends to fade over time, but this has no effect on their serviceablity. Provided none of the tiles are broken, a tile roof is a very long-lasting covering.

Slate

Stone has been in service as a roofing material longer than any other material, and is both incredibly durable and incredibly expensive. A slate roof will last almost indefinitely. The only obvious warning signs to look for are broken or missing sections of the stone tiles.

Flat Roofs

Although most homes have sloped roofs with some form of shingle or tile surface that is designed to shed water, some are built with flat roofs. And many homes with sloped roofs have a secondary section of roof—over a porch, garage, or entryway, for instance—that is flat.

Flat roofs cannot use traditional shingles, since water won't flow off them. Flat roofing systems generally come in one of two forms.

Built-up roofs are the older style of flat roof. They consist of layers of roofing felt (tar paper) and asphalt with an aggregate surface of gravel or crushed rock. Built-up flat roofs are hard to inspect, but be on the lookout for areas that have clearly been patched, or where there appear to be bubbles.

Membrane roofs consist of a layer of thick synthetic rubber attached with nails and roofing cement. Inspect a membrane roof for low spots where water can collect; in these areas, the membrane can break down and allow water to enter the home. In the example on page 47, the area where water is pooled was the entry point

This cedar shingle roof has weathered to an attractive gray, but is in great shape and will provide decades of service.

Metal roofs are experiencing a new surge in popularity. They are best suited for rustic home styles.

A clay or cement tile roof with no obvious signs of damage will be virtually trouble free.

A slate roof is elegant and extremely durable.

A shingle roof with irregular trim lines along valley flashing or end gables indicates bad workmanship. Because a roof is such a critical component, inspect any suspect roof job very carefully. Once inside the house, look hard for signs of water damage.

Vinyl gutter systems are very inexpensive and easy to install, making them a favorite of do-it-yourselfers. Make sure they are solidly anchored and installed with the correct downward slope to the downspouts.

for a major leak that ruined large sections of wall and carpeting.

Any flat roofing system should have some means of drainage—either a slight slope running to an overhang, or a center or edge drain opening. Beware of any flat roof where water pools. If you're considering a home with large expanses of flat roof, I advise that you hire a roofer to give you a detailed opinion of its condition.

Roll Roofing

Because water flows slowly off a slightly sloped roof, building codes require the use of roll roofing in these cases. It is used mainly on the roofs of porches or attached garages. Roll roofing is essentially granular shingling material sold in large rolls rather than individual shingles. Look for blistered or cracked areas and gaps along the lapped joints, where water might enter.

Roof Venting

To prevent moisture and heat buildup in the unfinished attic spaces below the roof surface, all roofs should have adequate venting. Without proper venting, moist air can cause building

Without proper venting, moist air can cause sheathing to decay. If this has occurred, you might see waviness in the overall surface of the roof.

Built-up roofing is rarely installed these days. Typical life expectancy is 25 years; replacement costs are about $3.15 per square foot.

This membrane roof is in good shape. Because this is a relatively new roofing system, its life expectancy is not yet known.

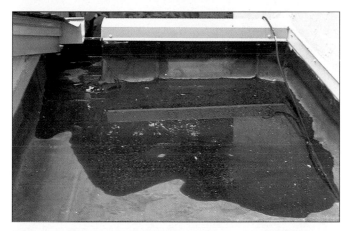

This membrane roof allows water to pool, which led to a major leak. Membrane roofs typically cost about $4.15 per square foot to install.

Roll roofing is required on roofs with slight slopes. Many porch roofs have traditional shingles, but should have roll roofing.

materials, especially sheathing, to decay. If this has occurred, you might see a waviness in the overall surface of the roof. In warm climates, an unventilated attic can make the entire house unbearably hot.

In addition, in areas with snowy winters, improperly vented roofs can cause ice dams. Ice dams occur when warm air in the attic melts snow, which then runs down the roof and refreezes as it reaches the cold eave overhang. Ice dams can back up under shingles and cause substantial water damage inside the home, so I always take this seriously.

Building codes may vary, but in general, an attic should have about 1 square foot of vent space for each 150 square feet of unheated attic space. For example, in a basic rambler-style house with a 30' × 50' foundation footprint, the attic space would be 1500 square feet, which would require 10 sqare feet of vents. Roofs can be vented in four different ways.

Roof vents are the familiar vents that you see on most shingled roofs. If this is the method used, there should be at least three such vents on any roof—more on an expansive roof.

A ridge vent is a shingle-covered raised strip that runs along the spine of the house. It

has vent openings along each side to provide continuous ventilation along the entire spine of the house.

Gable vents are slatted or gridded openings that are found on the flat gable end of the home's walls, usually near the peak. There should be one gable vent at each end, providing cross ventilation. Gable vents alone rarely provide enough ventilation and should be augmented by additional ridge vents or roof vents.

Soffit vents are metal panels installed in covered eaves to create ventilation in the overhang areas. Be suspicious if the home you're inspecting doesn't have vents in the covered eaves.

If you see no venting at all or if the vents are inadequate, this is a warning sign. All vents should have screens covering the openings to prevent insects and other pests from entering the home. On gable vents, you may be able to determine this using binoculars from the ground. When you get into the attic from inside the house, have a close look at the vents.

This roof has low lying vents and a ridge vent along the spine of the house.

Gable vents usually are louvered openings positioned on the vertical end walls of an attic space.

Soffit vents are installed in the overhang panels to circulate air into the spaces between roof rafters.

Turbine vents are roof mounted. Rising hot air from the attic, or exterior breezes, cause the vents to spin, facilitating the exhaust of heat.

Shown cut away

This staged model shows how ice dams form as water melts on a badly ventilated roof, refreezes when it hits the overhang, and backs up under shingles.

This is actual water damage caused when an ice dam backed up under shingles and melted into the living space below.

Flashing

Flashing is the pieces of folded or creased metal that form the seal between the roof line and intersecting surfaces, such as chimneys, dormers or skylights, or along the valleys between planes of the roof. Damaged or worn-out flashing is probably the most common location for leaking, so pay close attention to these areas.

With binoculars or by eye, look for rusted or patched areas where leaking might occur. Any gap between the roof surface and intersecting materials is a potential trouble spot. Ideally, you should see shiny new flashing with no obvious gaps of any kind. Old, corroded flashing often indicates an old roof, so if you see this sign, be on the lookout for other signs of roof deterioration.

A special type of flashing, called a *cricket*, is essential on any brick chimney that is positioned below the peak of the roof. A cricket is a wedge-shaped piece of flashing that diverts water running down the roof to the sides of the chimney, rather than allowing it to splash

From the outside, this house appeared to be adequately vented, which made the obvious ice dams hard to explain. From the attic, though, the explanation became obvious. During a recent reroofing job, the roofers had failed to line up the vents with the existing openings cut into the sheathing. This dramatically reduced the effectiveness of the vents.

Shiny new flashing (above, left) is a sign of a good, serviceable roof, while rusted or corroded flashing (above, center and right) is a sign a roof is old and may be leaking.

A chimney with a proper cricket (top) and one without (bottom). From inside the home, always check the ceilings around chimneys for signs of water damage.

directly on the side of the chimney and puddle there. Missing cricket flashing is a major cause of roof leaks, so if the chimney lacks flashing, you'd be wise to look for signs of water damage inside the house.

Now it's time to move your attention downward to the storm gutters, if the house is equipped with them.

Gutters

Storm gutters are a topic for debate. Properly sized, installed, and maintained, they direct runoff water from the roof well away from the foundation so that it can't infiltrate the basement. However, when they are not properly installed and maintained (which is often) gutters cause more problems than they solve. Clogged with soggy leaves, gutters can cause the wood fascia boards to rot. If the downspouts aren't properly installed, gutters can actually focus water on narrow portions of the foundation, causing more water damage.

As a result, I tend to judge a home more harshly for badly maintained gutters than for having no gutters at all. A home without gutters may have no water problems whatsoever, provided the grading around the home is adequately sloped away from the foundation.

If the home does have storm gutters, check to see if they are properly sloped toward the downspouts, if the downspouts have runoff pipes to carry water well away from the building, and if the seams are well sealed and show no signs of corrosion. Mesh covers over the gutters can help prevent leaves from clogging

This isn't a container garden, but roof gutters left unattended so long that a veritable urban forest has taken root. Trapped moisture may have already caused rot in the fascia boards and corrosion to the gutters.

them. Without covers, gutters need to be cleaned regularly. I've seen many unattended gutter systems that have tree seedlings sprouting from the composted leaf material clogging them, as in the photo shown here.

Vinyl gutter systems became enormously popular a few years ago, because the materials are so easy to cut, fit, and hang. This made them very common for do-it-yourself installations. Unfortunately, they're not very long-lived, and I generally don't view vinyl gutters with very much optimism. A better sign is the presence of seamless gutters. They tend to be a bit more expensive because they can only installed by a contractor with special equipment, but seamless gutters have a good long life expectancy that makes them worth the investment.

Unless they have been in place long enough to cause rot to fascia boards, bad gutters are more of a nuisance than a serious problem. Replacing gutters is a bit time consuming, but not a big deal, so never turn down a house strictly on the basis of bad gutters.

Gutters damaged in this way are a decided liability. Having no gutters at all is better than damaged gutters.

Signs of Gold:

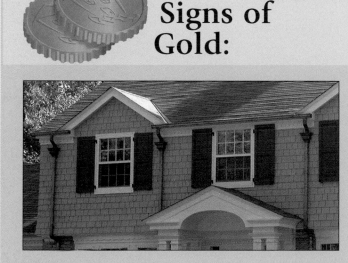

Copper gutters last a very long time, and indicate an owner who spared no expense.

Regarding Roofs: Inspection Checklist

Item	Good Indications	Good	Average	Poor
		Condition		
Ridge board	• Straight, minimal sagging	☐	☐	☐
Roofing material	• Granular shingles, new • Wood shingles, good shape • Tile or slate	☐ ☐ ☐	☐ ☐ ☐	☐ ☐ ☐
Venting	• Ridge vent • 1 sq. ft. vent for every 150 sq. ft of unheated attic	☐ ☐	☐ ☐	☐ ☐
Flashings	• New, no sign of corrosion • Chimney cricket flashing	☐ ☐	☐ ☐	☐ ☐
Gutters	• Seamless, good condition • Downspouts secure, extensions present	☐ ☐	☐ ☐	☐ ☐

If You're Selling:

If you have an old roof, you might wonder if it makes sense to have your house reroofed in order to make it more saleable.

Unless the roof is actively leaking, the answer is probably not. The cost of a new roof doesn't translate into a comparable increase in sale value, so it rarely makes sense to replace the roof when you're selling.

There are two exceptions. First, if the roof is actively leaking, you have both a legal and moral obligation to get things sealed up for the next owner of your home. If the appraisal inspection that lenders require spots the problem, you'll need to replace the roof anyway, so you might as well be proactive about this.

The second exception is if the real estate market is so soft that you'll need every possible advantage to sell your house. In this case, a brand-new roof instead of a 15-year-old roof might be the thing that puts you over the top.

Rooftop Anatomy

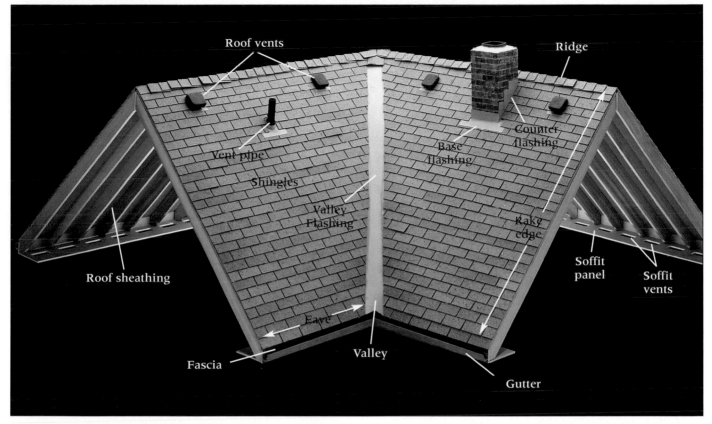

The elements of a roof system work together to provide shelter, drainage, and ventilation. Inspect as many of these components as you can during your review of the roof.

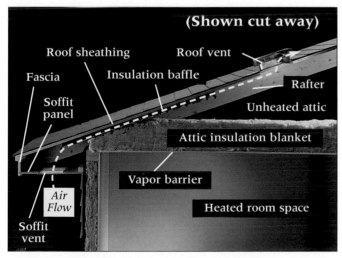

Good air flow prevents heat buildup and prevents roof damage caused by condensation or ice. Experts say the best ventilation is provided by a combination of soffit vents and roof or ridge vents.

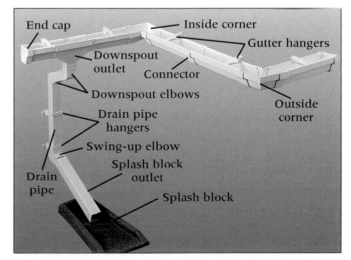

Vinyl or aluminum gutter systems have these components. Installing new gutters isn't very expensive, so don't worry too much if the home doesn't have gutters.

Chapter 7

Pick a Side

Assessing the Siding

Different siding materials may be used on the same house. This house has brick, painted wood shingles, and even areas of stucco beneath the windows.

Direct your attention down from the roof and begin to examine the siding material, another critical component of any home.

The siding forms the skin of a house, and its function is to protect the interior against infiltration by the elements—especially water. There are many different types of siding material commonly in use, and they all have unique merits and drawbacks. Signs of serious water damage are a red warning flag for any type of siding. Extensive water damage is one of the few flat-out deal killers you might run into. At the very least, the presence of this damage should be something you firmly address when it comes time to negotiate a price for the house.

LIFE EXPECTANCY AND COST OF COMMON SIDING REPAIRS

Siding Material	Life Expectancy	Replacement Cost
Scrape & paint wood siding	5 to 8 years	$1.25 per sq. ft.
Install wood lap siding	20 to 25 years	$3.50-$5.00 per sq. ft.
Install wood shingles, shakes	25 to 30 years	$4.50-$5.50 sq. ft
Install aluminum, vinyl siding	30 to 40 years	$3.50 per sq. ft.
Tuckpoint brick	10 to 20 years	$3.15 per sq. ft.
Replace brick veneer	60 to 100 years	$17.00 per sq. ft.
Install new stucco	40 to 75 years	$37.00 per sq. yd.

Inspect wood lap siding around openings for signs of decay (left), and look for separated joints (right) where water might have infiltrated.

Wood Lap Siding

This is by far the most common type of siding across the country. Wood lap siding is usually installed as beveled, overlapping planks placed horizontally. Its surface can be either smooth or textured. Cedar is the most common wood used, due to its natural resistance to decay. If it's kept well sealed or painted, cedar siding sometimes lasts as long as 40 years, though 25 years is generally regarded as the average life expectancy. It's often painted or stained, but rough-finished cedar can also be left unfinished to weather to an attractive gray color.

Focus your inspection on the bottom few rows of siding, and the areas under eaves and around windows and doors—wherever there's a likelihood of water penetration. Use an awl or small screwdriver to probe the wood and look for soft spots that signal wood decay. If you find signs of extensive decay, this is a fairly serious problem that could affect your decision to buy the home. The south and southwest faces of the home are most likely to have this damage, so focus your attention here.

Carefully study any points where plumbing

Tracks of the Poor Craftsman

Wood lap siding generally is installed in long planks with offsetting vertical seams. The presence of many short lengths of siding indicates that it was probably installed by someone trying to make use of small offcast pieces rather than spend the money on long planks.

Shingle siding with missing or loose pieces may be on its last legs.

Vertical siding, often called board and batten has a rustic appeal, but is not as weatherproof as overlapping wood siding.

pipes, vent pipes or other utilities enter the house. These areas should be caulked or sealed, and there should be no soft wood around them. If there is, it's a sign that water has infiltrated the siding.

Shakes or Shingles

Wood siding applied in an overlapping "fish-scale" style is likely cedar or redwood shingles (relatively smooth, machine-sawed pieces) or shakes (rough-hewn, hand-split pieces). This is a high-quality siding material that you should take as a good sign, provided it's in good shape. Generally, shakes or shingles are left unpainted to weather naturally to a neutral gray color. They can also be sealed or stained to help them retain their warm brown color, which many people like.

Look for signs of curling at the edges of the shakes or shingles, or severe splits. This condition tells you that the wood is badly aged. Walk around the house and gently pull at the bottoms of a few shingles. They should be firmly attached. If they "give" slightly, it means that the shingles may have been improperly nailed to the sheathing rather than attached to nailing strips that are anchored to the sheathing, as they should be.

Vertical Wood Siding

Most vertical wood siding is a board-and-batten style in which wide boards are spaced with noticeable seams that are covered by narrower boards. Look for large cracks or knots that have come loose. These problems can allow water to penetrate behind the sheathing. Vertical siding doesn't shed water as effectively as horizontal lap siding, so it's wise to study this surface carefully.

Exterior Plywood Siding

Exterior plywood panels are constructed with grooved surfaces to make this siding look like vertical wood siding, but plywood siding has long horizontal seams running between panels.

Plywood siding is less expensive and less durable than other forms of wood siding, so you should examine the surfaces carefully. Along the horizontal seams between the panels, Z-flashing should be installed to ensure that water doesn't penetrate the seams. If the Z-flashing is bent, or missing altogether, view the house with some suspicion.

Also look along the faces of the house from the side to see if any of the plywood panels are badly bowed. In some cases, the bowing can even pull nails out of the wall.

Plywood siding characteristically has vertical grooves. It is an inexpensive siding that is less desirable than traditional lap siding because it is subject to moisture damage, especially at the horizontal seams between panels.

Deal Killers:

Hardboard composite siding is to exterior siding what the hot dog is to a fine meats counter: not very nutritious or satisfying. Made from a porridge of saw dust, wood fibers, and glue, hardboard siding was developed as a low-cost alternative to wood siding. Because it is so unreliable, hardboard siding is no longer sold, but there are many thousands of homes that have this siding, and it's entirely possible you'll run across one in your search. If you find a house with this kind of siding, it's best to just assume that you'll be re-siding it at some point—probably sooner rather than later.

Although it comes in several forms, the most common form of hardboard siding is in long horizontal boards with a bead along the bottom edge. If you stand well back from hardboard siding and look at the surface from the side, you may see bubbled or blistered areas that interrupt the smooth surface. This is a sign of water damage and indicates that the entire surface is suspect. Walk around the house and press the siding at various points. Water-damaged hardboard siding feels soft. Pay particular attention to seams between boards, nail heads, and other openings where water can easily enter. Sometimes you'll see swelling where the hardboard has absorbed water.

Hardboard composite siding can be identified by the swelling that occurs at the ends, at the joints, and around nail heads.

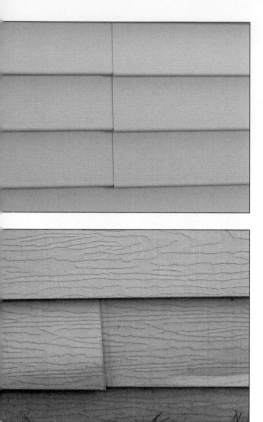

Vinyl (top) and metal siding (bottom) both have overlapping joints where the pieces fit together, and both may have a textured surface that resembles wood. They can be distinguished by tapping on the surface: metal siding will sound metallic.

Like vinyl siding, aluminum isn't perfectly weatherproof, so you'll want to probe around seams and other openings for signs that moisture is causing decay to the wood behind the siding.

Vinyl Siding

Vinyl siding is easy to spot. It often has a textured surface that's designed to resemble wood grain, but it has a distinctly "plastic" feel and has vertical overlapping seams where the sections of siding meet.

Assuming that it's not covering up some kind of serious rot, vinyl has distinct advantages as a siding material. It requires little in the way of maintenance, and properly installed and maintained, can last for 50 years or more. It's not waterproof, however, so just because the siding is vinyl you can't assume there has been no water damage to the surfaces behind the siding. In some cases, vinyl siding is installed directly over deteriorated wood siding—a bad idea. Pay attention to joints and cracks between the pieces of vinyl, and look for any signs that the surface has been punctured—by a hard-hit baseball, for example, or a falling tree branch. Gently probe with your awl or screwdriver around any openings to feel for softened sheathing beneath, which indicates that water has been a habitual intruder. (Make sure, though, not to damage any caulking or seals when you inspect these areas.) In winter, you might even see ice running behind the siding—another sign that water is a serious problem.

Vinyl siding systems usually are installed with vinyl window and door trim, and vinyl soffit panels, so look at these areas closely as well for places where water might enter.

Metal Siding

Like vinyl, metal siding at first glance can look like lap wood siding. When you look closely, however, you'll see that there are vertical seams where the panels overlap. Older metal siding often has a smooth surface; newer varieties may be textured to make them look more like wood siding.

Metal siding is usually aluminum, or less frequently, steel. Like vinyl siding, metal isn't perfectly weatherproof, so you'll want to probe around seams and other openings for signs that moisture is causing decay to the wood behind the siding—taking care not to damage caulking or weatherproofing. Aluminum siding can also be dented by hail, tree branches, or any sharp impact, so study the surfaces from the side to spot dings and dents.

Brick Siding

Brick siding is very durable and attractive, provided the mortar joints have been maintained. A few small cracks aren't a reason to run from a home, but wide cracks that have been in place for a long time can allow moisture to get behind the brick and cause extensive damage.

Deal Killers:

Long, continuous Z-cracks are a bad sign in brickwork. They not only allow water to penetrate, but can indicate a serious settling problem in the foundation. When you see cracks above window or door frames, it's a symptom that foundation settling has been uneven. This might well be a deal killer.

Obvious wide cracks in a brick façade are cause for worry, but so is the presence of obviously new mortar. New mortar suggests that the homeowner has recently patched major cracks to prepare the house for sale. If these cracks were present for a long period before they were sealed, the patches might hide long-term water damage.

In very old brickwork, you might see "rotting" bricks—bricks whose facings have crumbled away, leaving a rough, mottled appearance. A few bad bricks can be removed and replaced, but extensive rotting tells you that very, very expensive repairs might be required in the near future. For most people, this is a deal killer unless the asking price reflects the situation realistically.

Deformities in brickwork are sometimes quite obvious (top). Less dramatic but just as important are the presence of long Z-cracks in a new brick home (bottom). Both are symptoms of foundation settling. The top example is a sure deal killer—the bottom one requires an engineer's opinion to determine the severity of the problem.

Beautiful and deadly. If you choose to leave vines growing on a wood sided house, resign yourself to the fact that it will deteriorate faster than a bare wall.

Sidings vs. Ivy

Green growing vines such as ivy can be very nice to look at—and very hard on siding materials, especially wood. If you run into a house covered with ivy, take pains to look at the condition of the siding under the greenery, especially around window and door trim. Vines can trap moisture against a wall and hasten wood decay and disintigration of mortar, especially in climates that have long wet seasons. Removing a blanket of ivy from a house can be very, very time-consuming, but it's the best solution in the long run. If you like the look so well that you leave the vines in place (I'm guilty of this, I must confess), then you'll need to recognize that the siding will probably not last as long.

Cementitious Siding

Another siding material you might run into is cement tile, which sometimes contains asbestos. Don't worry: there is no health hazard to cementitious siding that contains asbestos. In fact, these siding materials are very durable. In the example shown here, even though the paint has long since worn away, there is no damage whatsoever to the siding.

Cementitious siding is a very brittle surface, however, and it's very common to see some type of damage. Cementitious tile isn't installed much anymore, so you may have difficulty getting damaged surfaces repaired.

Stucco

In most instances, stucco is an excellent siding material that is durable, offers good protection against the elements, and is easy to maintain. In a few instances, though, stucco can be a truly awful material.

Stucco is essentially a concrete surface that has been built up in layers. The final layer is tinted and applied so it has an attractive texture. A classic stucco surface is a very durable, impenetrable surface that can't be beat.

Cementitious siding is durable but brittle. If it is free of cracks and chips, it will last a long time.

Stucco can be repaired—but seeing discolored sheathing tells you that this wall has been open to water for quite some time. Check carefully to determine the extent of water damage.

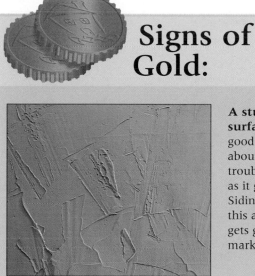

Signs of Gold:

A stucco surface in good shape is about as trouble-free as it gets. Siding like this always gets good marks.

Deal Killers:

Now for the awful part. There can be severe problems with a newer type of stucco installation commonly known as synthetic stucco. The technical term is EIFS (Exterior Insulation and Finish System). This type of stucco consists of a layer of rigid foam glued or nailed to a wooden backer board, which supports a layer of fiberglass mesh, a base coat, and a finish coat. The problem is that the surface is too watertight. When water does creep around window and door frames—some seepage is inevitable in any kind of wall—it becomes trapped inside the wall with no way to escape. The result is wood rot that can progress to a devastating degree before it's spotted.

It's difficult to identify a synthetic stucco installation with problems. The water damage can be virtually invisible until you pull off interior wall surfaces, and this isn't something you can do during an open-house viewing. You can, however, look carefully for areas where the stucco is cracked. You may be able to see something about the anatomy of the wall at these points. Look for the presence of fiberglass mesh—the hallmark of a synthetic stucco installation.

Synthetic stucco has become one of the biggest scandals in the building industry. Thousands of consumers have entered into class-action lawsuits against installers of this material. If you run across a newer stucco house, ask questions, frequently and often, about how it was installed. It might also be smart to hire a stucco contractor to take a look at the house before you sign a purchase agreement.

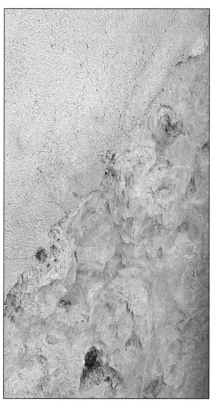

If a stucco surface is damaged, you may spot tell-tale signs of the fiberglass underlayment that signals synthetic stucco. If you suspect synthetic stucco, hire a stucco contractor to evaluate it.

Stone Facade

Unless you happen to be inspecting a castle, most any home that appears to have solid stone walls is actually built with a stone veneer laid over a wood framed or masonry wall. Evaluate these walls in much the same way you'd evaluate brick: look for bad cracks that may have exposed the interior of the walls to long-term moisture. Also look for areas where the veneer has pulled away from the framing—sometimes you can hear a hollow sound if you tap lightly on suspect areas of the walls.

Look carefully around windows and doors, since this is where a stone facade is most likely to have gaps that allow moisture to enter.

A stone facade siding is a top-quality surface found only in quality homes. As with any siding, look closely for cracks and signs of water damage (right).

Problems with stone facades are most apparent where the stone meets other surfaces—in this case, the metal support lintel that frames a picture window. You should be a bit concerned about water damage if you see a stone or brick house with many gaps around the windows and doors.

If You're Selling:

It almost never makes financial sense to re-side a house in order to sell it, because the cost of the new siding rarely translates into an equal increase in sale price. It's better to simply be up front about any problems with your home's siding and establish an asking price that takes this into account.

It does make sense, however, to give your home's exterior a fresh cleaning or coat of paint. If you don't have the time to paint the entire house, pay particular attention to wood trim moldings, making sure they are painted and caulked. Conservative paint schemes are best, and coordinating the trim so it complements the color of the roof is a good idea.

Pick a Side: Inspection Checklist

Item	Good Indications	Good	Average	Poor
		Condition		
Age of siding	• At least half life remaining	☐	☐	☐
Surface condition	• Solid, gaps caulked	☐	☐	☐
	• Firm, no sponginess in siding or sheathing	☐	☐	☐
Paint or stain surface	• Bright, intact	☐	☐	☐
Insect or pest damage	• No insect trails, bird holes	☐	☐	☐

Chapter 8

Step Inside
Checking the Layout

© Karen Melvin

When I go indoors after looking over the outside of a house, my first step is to get a feel for the general layout of the place. Although it is, of course, possible to change the size and organization of the rooms through major remodeling, this is awfully expensive. So evaluating the layout of a house is an important step in deciding whether it warrants closer inspection.

Don't rush this part. The initial walk-through can be telling, and your intuition means a lot. Does the layout feel wrong or inconvenient, or does it feel natural and inviting? It is important to use to your imagination. While walking through the house, mentally run through a typical day in your life. For example, if your household includes kids, can you visualize your entire family getting ready for work and school in the morning? Would the kitchen and dining room comfortably hold a typical family dinner with a few guests joining you? Also think about those less typical days—a big holiday celebration or the occasional cocktail party. Does the house accommodate your lifestyle, or would your lifestyle need to accommodate the house?

Beginning at the entryway, move through the house following logical pathways: entryway to living areas, living areas to and from the kitchen, bedrooms to bathrooms, and so forth. Take special note of the transition spaces: steps between levels, staircases, and hallways. Are there any built-in traffic jams or accessibility concerns? Does the house feel chopped up and confining? Open, but impersonal and lacking in privacy? Spacious but intimate? Flexible and efficient? Keep these and other words in mind as you walk through, and listen to your intuition. A truly negative first impression is enough to send you on to the next house, but a positive one tells you it's time to dig deeper.

Signs of Gold: A finished attic or basement, if it's well designed, creates a lot of value in the home, and maximizes use of space.

The Individual Rooms

After the initial walk-through, evaluate individual rooms and spaces. Room sizes can vary greatly and are largely a matter of personal taste, but it's good to have some goals in mind. If your current home or apartment has certain rooms that function well for you, compare the rooms in the new house to see if they are comparable in size. If your goal is more space, look for rooms that are larger than you currently have; if more intimate spaces are desirable, you might prefer rooms that are smaller. Be sure to note the locations of receptacles, light fixtures, windows, and doors. Your digital camera can be helpful for recording details of specific rooms.

Bathrooms. The bathrooms are crucial to the everyday livability of a house, so it's wise to evaluate them first. Inadequate or poorly designed bathrooms can be a deal breaker, because extensive remodeling and expansion of these spaces can be very expensive.

In bathrooms, size matters almost as much as number. Five feet by 8 feet is a minimum for a primary family bath. Half baths can be—and often are—quite small, and it's important that they be well designed. If a half bath is so small that you can't turn around in it without hitting something, it won't be comfortable or usable.

Finally, bathrooms are like kitchens in that they must be functional. A beautiful bathroom is great, but its layout must accommodate the ways you'll use it. If more than one family member uses a bathroom at the same time while getting ready in the morning, it's a good idea to imagine or even actually go through the motions together. Is there enough counter space? Storage? Sink space? A huge whirlpool tub may be just the ticket, or may be a wasted luxury if your family prefers showers.

Bedrooms. Having too few—or too many—bedrooms is a common reason for shopping for a new house, so make sure the house you're inspecting has what you need. Remember that the distinction between a bedroom and den or study can be a fine one. Here are some things to consider with rooms you're envisioning as bedrooms:

- Bedrooms should not be smaller than 90 square feet and closets should be at least 4 feet by 2 feet.

Signs of Gold:

Half baths like this one carved into otherwise wasted space can be a real plus. Toilets and sinks should have at least 21 inches of clear walkway in front. Doors should swing out if they cannot swing in freely.

By imagining how you would function in the house and how your things would fit into it, you'll get a better sense of how the house will work for you.

Egress windows provide an emergency escape and are required by code for any attic or basement bedroom.

Kitchens are above all else working spaces, and any type of kitchen should reflect this in its layout.

- Appliance noise can penetrate poorly soundproofed walls.
- Well-placed closets make good noise barriers.
- Bedrooms on the front wall of a house may be subject to more noise from the street.
- Orientation of the house: are you a morning person who likes sun streaming in during the morning, or a night owl who wants things dark until late morning?
- Basement and attic rooms converted to guest rooms or bedrooms must always have adequate egress windows.
- Bedrooms should have convenient access to bathrooms.
- A master suite is more private if it's located well away from children's rooms or entertainment areas.

Kitchens. While bathrooms and bedrooms are important, practical concerns for home buyers, the kitchen is often the showpiece of a home—especially a newer home—and will likely play a large role in your buying decision. I know many families for whom the kitchen is the true family gathering place and the most important room in the house. A minimum size for such a kitchen should be at least 110 square feet, but today it is not uncommon for the kitchen to rival the family room in size. Keep these things in mind as you imagine using the kitchen:

- Most families prefer an open kitchen plan with an island or breakfast nook.
- Any kitchen should adjoin the dining room or the family room and any other areas where you frequently eat.
- Any kitchen should have ample and convenient storage space.

The traditional kitchen is—above all else—a working space, and this should be reflected in its layout. You can get a feel for this by measuring the "work triangle"—the geometric figure with the range, the sink area, and the refrigerator at its three points. The sides of the triangle should not intersect with any high foot-traffic area, and the length of any of the sides should be between 4 and 9 feet and the total perimeter should be between 12 and 26 feet—anything less suggests a cramped workspace. In addition, a functional kitchen should have at least one stretch of clear countertop space that is at least 4 linear feet long.

Today's kitchen, though, just as often serves as a snack shop where family and friends gather for informal meals and fun. A kitchen with a booth-style seating area or peninsula countertop seating area is very appealing to many people. Some kitchens even have small secondary refrigerators for storing snacks and refreshments.

A kitchen is the most expensive room in the home to remodel, with prices starting at $5,000 for a relatively superficial remodeling, and easily ranging up to $40,000 to $60,000 if the kitchen requires expansion or a completely new layout. If the kitchen is a crucial space for you, a substandard kitchen might well be a deal breaker.

Finally, consider the solar exposure of the kitchen and the number of windows. A kitchen with few windows and a northern exposure could be dreary and despressing, while one with huge patio doors and picture windows facing south can be uncomfortably hot on a bright day when the cooking appliances are in full use.

Living and entertaining areas. Ideally, designated living and family rooms are large, open, main-floor areas, although in many homes these spaces are now found in basements. Beyond square area (a minimum for a living room is 150 square feet; 220 square feet for a family room), your taste, furnishings, and equipment will dictate the suitability of living and entertaining areas. A small living room with narrow doors will not convert into an entertainment room with a billiard table, though it might be pleasing if your family prefers smaller spaces for entertaining. Ideally, a house will have both spacious and intimate living areas—not every living room needs cathedral ceilings.

Make sure to spend at least a few moments in these rooms taking measurements and imagining how furniture and equipment might fit. I know one buyer, a dedicated musician, who sold his new home within weeks because he hadn't checked to make sure the parlor would hold his classic grand piano.

It's always a good idea to walk through a house with everyone who will live there, but it's particlarly helpful when evaluating community spaces that the entire family will use.

Placement of these areas within the overall layout of the house is important. I try to imagine the room in use and how that use will affect the rest of the house. For instance, if a large room adjoins the foyer, it will likely be there that you greet guests to your home. That means it won't be the best space to

A sunny breakfast nook can improve a kitchen without taking up floor space. Kitchens especially benefit from clever use of space.

Signs of Gold:

While a spacious kitchen is almost always a sign of gold, a large, open kitchen with access to the dining room, the family room, and the outside (to a deck or where a deck could be) is a truly convenient, flexible, modern kitchen.

A house with room to grow is definitely a good thing. A house with unused room to expand will come cheaper than a fully finished house, while allowing you to grow the house to fit your family as needed.

A house that takes advantage of Universal Design and accessibility principals in its layout—even if you don't appreciate it immediately—will be more accomodating as you age and more valuable should you sell.

Built-in storage features, like this dresser recessed into an attic knee-wall, can make a home seem larger by eliminating the need for bulky dressers and cabinets.

use as a family recreation room with the attendant televisions, videogames, and clutter. I also try to imagine how noise from these high-traffic rooms could spread through walls or floors into other rooms—especially bedrooms. Test noise leakage whenever you can, and determine where the noise you hear is coming from. Remember, a kitchen or a laundry room can be noisy even when no one is in it.

Storage areas, closets, & pantries. Little things can be crucial to a successful home layout. Even if a house has all the major rooms in the right places, you need to examine the storage areas, closets, pantry, and other small, multipurpose spaces. Think about the places where you expect to find storage. Look for storage that seems intuitive, and ask yourself if the house is adaptable to your habits and needs. Find the closets and imagine what might go in them—and remember, closets are flexible spaces, so don't be limited by how the current homeowner is using them.

Utility spaces. Although they're not glamorous, the laundry, garage, and workshop can be critical spaces. Check out these areas and make sure they give you ample space for home maintenance, repair, and tool and supply storage. For a workshop or laundry, good lighting and a spacious working area are decided pluses.

If You're Selling:

Maximize your layout by removing as many furnishings as you can without making the house look empty.

Remember that buyers might not use rooms the same way you do, and if you can help them see potential, your house will be more marketable. The idea is to make the house simultaneously attractive and full of potential.

If you have planned a renovation (a screened-in porch or a larger garage), but have not started it, show the plans to potential buyers. It may help them see the home's potential and add to their interest.

Step Inside: Inspection Checklist

Item	Good Indications	Condition		
		Good	Average	Poor
Entryway	• Storage for outerwear	☐	☐	☐
	• Ample space for several guests	☐	☐	☐
Hallways	• Ample width	☐	☐	☐
	• Well lit	☐	☐	☐
	• Smooth traffic flow	☐	☐	☐
Stairs	• Spacious and safe	☐	☐	☐
	• Well lit	☐	☐	☐
Bathrooms	• Minimum 5 × 8 ft. in size	☐	☐	☐
	• Sufficient number	☐	☐	☐
	• Desired features	☐	☐	☐
	• Convenient half bath	☐	☐	☐
Bedrooms	• Sufficient number	☐	☐	☐
	• Ample size (90 sq. ft. minimum)	☐	☐	☐
	• Secondary escapes	☐	☐	☐
Kitchen	• Convenient to eating area	☐	☐	☐
	• 110 sq. ft. or more	☐	☐	☐
	• Efficient work space	☐	☐	☐
	• Good lighting	☐	☐	☐
Living/ entertaining areas	• 150 to 220 sq. ft. minimum	☐	☐	☐
	• Configured for desired furniture	☐	☐	☐
	• Good position within house	☐	☐	☐
Storage areas	• Sufficient closet space	☐	☐	☐
	• Built-in storage	☐	☐	☐
Utility spaces	• Well lit	☐	☐	☐
	• Ample work space	☐	☐	☐
	• Includes both laundry and workshop	☐	☐	☐

Chapter 9

From the Ground Up
Inspecting the Home's Structure

A good foundation is impervious to the effects of water. This concrete foundation has proper grading that directs water away from the wall, and the basement windows are constructed and landscaped in a way that ensures that water can't enter.

Major structural problems are among the few issues that can literally destroy a house, so make sure to pay close attention.

Somewhere near the start of your inspection, look very carefully for structual problems with the house. Major structural problems are among the few issues that can literally destroy a house, so pay close attention, and don't waste further time if you see ominous signs.

At my last serious home-buying excursion, my wife and I visited a home that had a peculiar feel as we walked through. It was hard to put a finger on it exactly, until I set a large marble on the hardwood floor in a spacious upstairs bedroom and watched the marble begin to roll, gathering speed until it ricocheted off the far wall. The home had settled in such an uneven fashion that it was quite evident to the senses—the marble merely proved what we felt. Although the real estate agent downplayed the issue, most everyone who looked at the house sensed the problem, because I watched the house remain on the market for many months. Eventually, a big rainstorm caused the low-side foundation wall to buckle inward in a way that couldn't be ignored. The wrecking ball came a week later.

Signs of structural problems can be visible both inside and outside the house, and sometimes in ways you might not expect, so don't be distracted by other things during this portion of the inspection. Professional inspectors take foundations very, very seriously; so should you.

Types of Foundations

The foundation on which your home rests will be one of three types: a *wall foundation*—either a full basement or a crawl space configuration; a *slab foundation*, which supports the house; or a *pier foundation*, where the house is supported by beams that rest on columns that extend into the earth. The type of foundation depends partly on the region in which you live, and partly on the age of the house. In most cold-weather regions, the houses you inspect will have wall foundations with basements or crawlspaces, while in

Wall foundations are made of stone, poured concrete, or concrete block, depending on the age and location of the house.

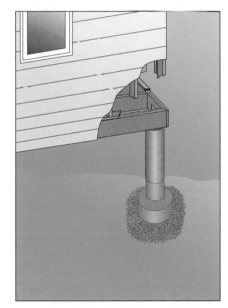

Pier or pile foundations are common in the Deep South and in some rural areas. In cold-weather areas, porches on older homes often rest on pier foundations.

Slab foundations are common in dry, warm-weather climates.

regions with warm weather or high water tables, most houses have slab or pier foundations.

Basement Wall Foundation

From the outside. Look first at the building site itself. If it's positioned on a hillside, this alone should put you on alert. A hillside that directs water toward one wall of the foundation can create enormous pressure in the soil. Look carefully at the outside of the foundation on the side that faces the hillside for signs of foundation cracks. When you get inside the house, very carefully study the foundation wall that faces the hillside. If there are soil pressure problems, this is where you'll see them.

While you're outside, look at the drainage around the house. Concrete slabs that have settled so they aim water toward the foundation are warning signs telling you to be on the lookout for basement water problems. Ideally, the landscape around the foundation should slope slightly away from the wall; if not, it's another warning sign. And look for the presence of gutter downspout extensions that carry rainwater well away from the foundation. Again, a lack of good downspout extensions tells you to look carefully when inside. It's possible that none of these problems has

Look first at the building site itself. If it's positioned on a hillside, this alone should put you on alert.

A tale of three downspouts. The first example (above) is a good installation—an extension diverts water a full 6 ft. away from the foundation. In the example at right, the presence of dirt on the sidewalk suggests that rainwater puddles there—a situation that could be dangerous in a freezing climate. In the last example (below) there is no downspout extension at all.

created meaningful water damage; even so, you'll want to correct them should you buy the house.

If the house has an exposed exterior stairway entry to the basement, look to see that there is a drain in the floor of the stairwell. If not, the stairwell might be a point where collected rainwater can enter the home, especially if the stairwell lies below a roof overhang rather than on the gable side of the home.

The interior walls. Now move inside to the basement. Begin by looking at the unfinished spaces, if there are any, because it's easier to judge things there than in a finished basement. The most important thing to look for is bowing or bulging in the stone, concrete, or block walls. If you find this, it means that soil outside the house has exerted enough pressure to cause the walls to begin to give. If the house is built on a hillside or slope, focus your attention on the uphill side, where most of the pressure will be located.

Not all cracks in a block or concrete wall are serious. It's perfectly normal to see a faint crack or two in a wall. A faint Z-shaped crack like the one shown on the opposite page is probably not a serious problem if it has opened over a period of several decades. Hold your face close to the wall, preferably with a strong sidelight, and look for signs that the wall bulges. This is a much more serious problem that might be enough to halt your inspection right on the spot.

Look for signs of white powder, called *efflorescence*, on concrete or block walls. Efflorescence occurs when water seeps through a masonry wall, evaporates, and leaves mineral residue behind. A masonry wall with extensive efflorescence is not only a wall that is letting water into the home, but is also at serious risk for structural collapse.

This stairwell has two problems: no floor drain to eliminate rainwater (left) and no railing (right), a condition that poses a decided safety hazard. Upon close inspection, the basement of this house showed interior water damage near the doorway.

Sump pump. Finally, look to see if the basement has a sump pump and drain tile system (see page 75). A drain tile system is a circular pathway of porous pipe installed beneath the floor of the basement around the perimeter near the wall. The pipe leads to a sump pit, fitted with an automatic electric water pump. When excess soil moisture running down the outside of the foundation wall reaches floor level, it enters the drain tile pipe and is diverted to the sump. The pump then activates and forces the water up and out of the house. If the system operates correctly, it prevents most moisture problems in the basement.

But if the pump has not been maintained correctly or was a recent addition, it's possible that you'll find evidence of water damage in the basement.

Pay particular attention to the wall areas near the floor. In finished spaces, look for discolored wallboard or baseboard moldings, as well as for the presence of mold on the walls and floor. A musty smell in the air indicates a chronically wet basement, as does the presence of an industrial strength dehumidifier.

Posts, Beams, & Rafters

Joists and foundation. If possible, closely inspect the points where the joists and support beam rest on the foundation. If you see gaps between these framing members and the foundation, it means that the entire foundation is settling, and that the house has begun to balance on the inner post footings rather than the outer

A faint Z-crack in a basement wall, like the one shown here, is fairly normal and is not a cause for concern. It is caused by minor soil settling.

Deal Killers:

This foundation wall faces a steep hillside (above, left), which should always make your radar kick in. Water draining down a hillside after a heavy rain can put enormous pressure on the soil next to the foundation wall. Once inside this home, I noticed that the wall facing the hillside had a large amount of white efflorescence (right, top), which indicated that lots of moisture has seeped through the wall. Even closer inspection might show that the wall developed bulging cracks that had continued to open up, even after they were patched with mortar (right, center).

The ultimate solution to a basement foundation that has suffered major stress due to soil pressure is a complete excavation and reconstruction of the wall (right, bottom), an enormously expensive proposition.

Signs of Gold:

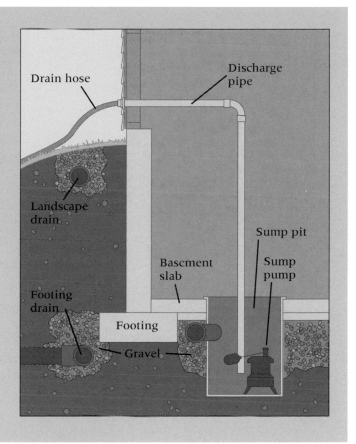

Discharge
pipe

Drain hose

Landscape
drain

Sump pit

Basement
slab

Sump
pump

Footing
drain

Footing

Gravel

A sump pump system that is properly installed and well maintained can keep a basement perfectly dry. To test a sump pump, remove the cover and pull up slightly on the float ball you find there. The pump should kick in. The house where this sump pump is located had a basement that was utterly dry and quite useable for any type of living space you might want.

Deal Killers: Even a sump pump can't cure severe water problems. This basement had two sump pumps, which ran constantly during my inspection on a clear, sunny day. And even with the pumps, the walls were covered with mold. A basement with very severe water problems will never be practical to use for real living space.

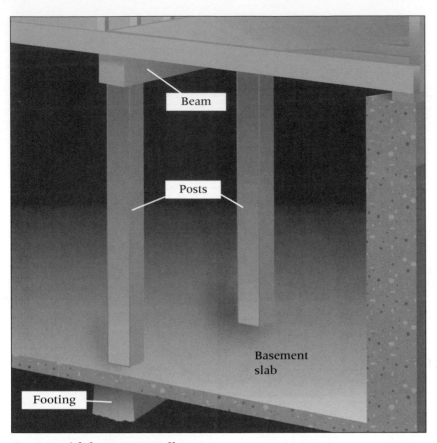

Homes with basement wall foundations also have at least one center support row of footings, posts, and beams.

Cracked or tilted posts offer no real structural support. Have an engineer examine any home with a questionable support structure.

foundation walls. This problem can be halted by shimming beneath the framing members to restore the bearing connection to the foundation, but it's possible that the damage has already been done. If you see this sign, look very closely at the condition of the walls and floors in the upper stories of the house.

Insect damage. While you're inspecting the sole plates and joists resting on the foundation, look for signs of decay and insect damage. In warm coastal areas, termites and other wood-eating insects can destroy homes. If you live in one of these areas, always have a suspect home inspected by an extermination expert or structural engineer before making an offer. Not all wood-eating insects are termites, though, so don't panic.

Generally speaking, the presence of carpenter ants or certain beetles isn't a deal-killing symptom. Carpenter ants generally feed on dead, decaying wood, and seeing them means you've found areas where there is wood decay. It's something to investigate, but it doesn't necessarily mean that insect damage has compromised the walls throughout the house, as may be the case with termites.

Center beam and posts. Now turn your attention to the posts and beams that support the center of the house. The support beams may be solid wood, layered beams made up of several pieces of dimension lumber, an engineered wood product, or even a steel I-beams. Look for any signs of bowing or cracking in the beams. The support posts may be solid wood timbers or metal posts. Check to make sure these posts are perfectly vertical and undamaged.

The presence of add-on posts tells you that the home has experienced some settling problems that the homeowner has addressed. The questions for you are how much structural damage occurred before the problem was fixed, and whether the settling is still occurring. If you see additional support posts, such as telescoping metal posts, be sure to ask questions of the homeowner.

Floor joists. Now turn your attention to exposed floor joists above your head. Inspect for cracks, which compromise the strength of these members. You should see diagonal cross-blocking between the joists, which helps to keep the floor above firm and rigid.

If you're inspecting with a partner, have him or her walk about on the floor above while you watch from below. If you can observe the joists visibly giving under foot traffic above, it may mean that the joists are undersized for the load they are carrying, which can result in a floor that permanently sags. If this is the case, you may see more obvious signs when you move upstairs.

Floor slab. Finally, have a look at the concrete floor slab. Look for signs of heaving or cracking. Again, faint cracks aren't a problem, but an offset crack that seriously bulges indicates that ground water surging up may be affecting the slab. And also look at the slab around floor drains. If the sewer periodically backs up, you may see water stains on the floor around these drains.

Inspecting Finished Basements

In fully finished basements it can be a bit harder to do a structural inspection, but here are some things to look for:

- Water stains along baseboards and lower walls. This may mean that the underlying foundation wall is allowing water into the home.
- Vinyl flooring seams that have loosened, and musty or damp carpeting, especially near the walls. These are signs of water damage.
- One or more dehumidifiers. All basements can be humid, but one that requires several powerful dehumidifiers may well be getting soil moisture as well as atmospheric moisture.
- Doors and windows that pinch or bind. If you see signs that they have been trimmed to make them fit, it's likely that the foundation has settled unevenly or has bowed due to soil pressure.

This is a carpenter ant nest, unpleasant, perhaps, but not nearly as dire as termite damage.

Severe termite damage can be identified by channels cut into the wood by feeding insects.

Look for cross-blocking between floor joists. It's a sign the entire house has been well-framed.

Photos courtesy of University of Nebraska Department of Entomology

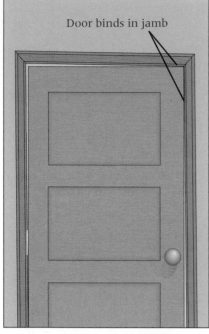

Door binds in jamb

Basement doors and windows that bind can be a sign that the foundation walls have settled unevenly.

Slab Foundations

If the home you're inspecting has a slab foundation, look for signs of heaving or cracking in the concrete slab on which the house is built. (Remember that homes with basements may also have garages, porches, or room additions that rest on slab foundations.) Generally, a heaving slab is caused by ground water rising from below, so you may find signs of water damage to flooring and the lower areas of walls. Again, faint cracks aren't a problem, but a crack that bulges indicates that ground water may be affecting the slab and the structure of the house. A slab foundation that is being damaged by rising water or settling soil will also announce itself by doors and windows that don't fit well in their frames. A heaving slab may also cause support posts in the center of the house to bow or crack.

Pier & Column Foundations

Pier and column foundations are made of stone, brick, concrete block, wood, or steel. In some parts of the country, the entire house rests on piers and columns. But it's also fairly common for an older house with a full basement wall foundation to have a porch that rests on piers and columns.

When you find pier and column foundations, look to make sure that all the columns are in firm contact with the support beams above. Stone columns often use shims or wood wedges; if the wedges

Deal Killers:

In a slab foundation, the sole plates and other framing members should be raised away from ground level to prevent rot. If you find a situation in which the wood structure rests directly on the ground, the structure is very likely to have experienced decay.

A slab foundation should be elevated at least 6" around the perimeter to lift the wooden framing members away from ground level. If the home construction hasn't followed the code recommendations, this can be the result: extensive rot to the sheathing and sole plates. It would take extensive work to repair this problem.

have fallen out, these columns may be offering no support to the structure above. If columns are mortared brick or block, make sure all the mortar joints are intact.

Watch for any signs of rot in wooden posts that rest directly on concrete. Decaying wood will be soft and crumbly. If the columns are brick, make sure that they are straight, and that the mortar joints are intact.

Where an attached porch is supported by piers and columns, study it carefully, because faulty columns can compromise the entire structure. A sure sign of a problem is when you see that the porch is beginning to separate from the main portion of the house. This indicates that the porch is settling and that the foundations are not secure. Rebuilding a porch can easily cost $10,000 to $15,000 or more, so pay attention when you see signs of trouble.

Pier and column foundations should be inspected carefully. Brick or stone columns should have mortar in good shape.

First Story

Now move from the basement to the first floor. Begin by looking at the floor for signs of bowing or settling. On hardwood or vinyl-covered floors, roll a large marble across the surface to see if rolls to one side, indicating that the floor slopes.

Look closely at the baseboard trim. If the small base-shoe molding has lifted away from the flooring, it may mean that the entire foundation is settling. In a home with this problem, you may actually be able to see the bow in the floor, crowning over the central support beams and sloping down toward the outside walls. This is a pretty common situation in older homes and is not a deal-breaking problem unless it is very pronounced.

Now inspect the walls. A few small cracks in plaster or wallboard aren't a big deal, but if you notice a pattern of cracks extending up from the corners of window and door frames, it can be a sign that the entire structure is suffering some uneven settling and shifting. This is a problem worth noting and investigating further. Before buying a house with such symptoms, it's wise to have it evaluated by an architect or structural engineer.

If the walls are plaster, tap randomly over the surfaces. If the walls have spots that sound hollow, it means that the plaster may be separating from the underlying lath. Repairing plaster walls is not difficult or expensive, but if you find many such areas, you

A gap between the shoe molding and the flooring may indicate that the foundation has settled.

Minor cracks in plaster are no big deal; view cracks more seriously if they extend from wall to ceiling, or if there is a pattern of cracking above windows and doors.

If the house has a second story and you're inspecting with a partner, have the partner go upstairs and tromp around a bit while you watch the ceiling.

could be in for a messy and time-consuming remodeling project. More to the point is the fact that serious plaster damage may indicate severe structural shifting, or that extensive water problems have affected wall surfaces.

Now look at the ceilings to see if they appear flat and level, and whether the surfaces are in good condition. In older homes, you may see areas where plaster is bowing. These will need to be repaired someday soon but aren't all that serious, in and of themselves. If the house has a second story and you're inspecting with a partner, have the partner go upstairs and tromp around a bit while you watch the ceiling. Can you see the ceiling give under foot traffic from above, or do ceiling fixtures jiggle and rattle under foot traffic? If so, the ceiling joists might be undersized for the load they carry—a condition that will cause the floors to sag over time and put steady stress on ceiling surfaces.

Second Story and Attic

If the house has a second story, and possibly a third, go upstairs and repeat the evaluations of the floors, walls, and ceilings. If the second story has angled ceilings that conform to the roofline, inspect this surface carefully for indications that nail and screw heads have rusted and bled through. This is a sign of inadequate insulation or ventilation, which may be leading to decay and destruction of the roof decking itself.

If there are skylights installed in the ceiling, study the edges of the skylight shaft for signs of moisture damage, which may suggest that the window flashings have failed and allowed moisture to penetrate the ceiling surface. The presence of moisture can also be caused simply by condensation of moist indoor air against the surface of the glass.

Finally, make every attempt to get into the attic or unfinished crawl spaces. In many modern houses, the attic space is constructed from framing trusses, and your only access to the space will be through a hatchway located in a hallway or closet. If possible, you should at the very least get on a stepladder and peer up into this space with your flashlight. Look for four things:

- Roof framing members that are sturdy and undamaged
- Roof sheathing that is dry, without any signs of decay
- An ample amount of ceiling insulation—generally at least 12"
- Adequate ventilation, through roof vents, end vents on the gable ends, or ridge vents running down house's spine

Possible Deal Killer

Vaulted ceilings in living rooms and great rooms are very popular in homes built over the last 20 years or so. These ceilings hold potential for some pretty major problems if the ceiling surfaces have been attached directly to roof rafters without proper insulation and, more importantly, proper ventilation. In a great many such homes, moist air rising through the ceiling can actually rot out roof decking, leading to many thousands of dollars in damages. Whenever you see a vaulted ceiling that rests against a sloped roofline, go outside to see if the house has both soffit vents and a ridge vent. This combination of venting offers the best safeguard against moisture buildup problems on vaulted ceilngs.

Sometimes you can spot signals that a problem exists by the presence of wallboard screws or nails that are rusting and bleeding through the surface finish. In a very tall room, I sometimes use binoculars to carefully inspect the ceiling surface to see if moisture has rusted screw or nailheads. If so, you'll need to investigate further.

In a walk-up attic, look for all these things, but also examine the areas around chimneys and vent stacks penetrating the roof. You're looking for signs of daylight peeking through, which indicate that the flashings may not be intact and that water could be entering the attic space.

If You're Selling:

It's a good investment to have an architectural engineer inspect your home and give you a written report identifying any problems and confirming the overall structual integrity. This is a specialized task that few standard inspectors are qualified to do, so make sure you've hired the right person for the job. This is an especially good idea if your home has some minor structural aberations that aren't particularly serious. A prospective buyer will be reassured by a report from an accredited engineer that says your minor foundation cracks are stable and not expanding.

In the attic, check around chimneys and vent pipes for signs of daylight—indications that the flashings aren't weather tight.

From the Ground Up: Structural Checklist

Item	Good Indications	Condition		
		Good	Average	Poor
Foundation	• Level, straight	☐	☐	☐
	• Dry	☐	☐	☐
	• Soil grade slopes away from foundation	☐	☐	☐
Downspout extensions	• Direct water away from foundation	☐	☐	☐
Window & stair wells	• Dry, drain present	☐	☐	☐
Interior foundation walls	• No sign of efflorescence	☐	☐	☐
	• Small cracks	☐	☐	☐
	• Finished walls are free of water damage	☐	☐	☐
Sump pump	• Operates correctly	☐	☐	☐
Posts, beams, rafters	• Straight, no damage	☐	☐	☐
	• No signs of insect damge	☐	☐	☐
	• Cross-blocking	☐	☐	☐
First, second stories	• Wall, ceiling cracks small or absent	☐	☐	☐
	• Level floors	☐	☐	☐
	• Windows, doors operate smoothly	☐	☐	☐
Attic	• Framing members intact	☐	☐	☐
	• Dry	☐	☐	☐

Anatomy of a Home's Structure

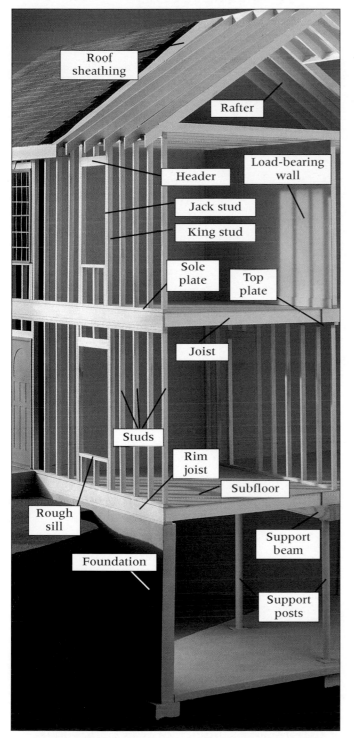

In **platform framing**, a house is constructed with a series of platforms and framed walls that support the roof framing.

A traditional rafter attic may be usable for storage space or as an expansion living space, provided the floor joists are large enough, and there is sufficient headroom. Look for $2 \times 10'$ or $2 \times 12'$ floor joists spaced 12″ or 16″ apart.

Roof trusses are found in the attics of many newer homes. These spaces cannot be used for storage or expansion living space. Make sure the framing members are all solidly anchored, without cracks or breaks.

Chapter 10

In, Out, and Through
Evaluating the Windows and Doors

Credit: Bill Tijerina Photography

Well-designed doors and windows add great curb appeal to a home. If well maintained, they should give years of good service.

Windows and doors are vital components in the appearance, safety, and climate control of a home.

They're way too often taken for granted, but windows and doors are vital components in the appearance, safety, and climate control of a home. We expect them to repel unwanted elements—from insects, pests, and drafty winds to human intruders—yet offer easy passage to family members, sunlight, pleasant breezes, and welcome guests. It is important not only to like how they look, but to evaluate their function and potential problems.

Inspecting the windows and doors requires that you study them from both inside and outside the house, and that you put them through their paces by opening, closing, locking, and unlocking them. In your evaluation of windows and doors, don't forget to look at the garage doors and windows and at any skylights that might be present.

I always look carefully for serious problems with windows and doors—the type of thing that will require replacement. But I also watch for pesky problems that might add up to prohibitive expenses. Since the average house has from 15 to 30 windows, two to eight exterior doors, and any number of interior doors, even minor issues may gradually nickel and dime you into spending a small fortune. Of course, if you're handy and can do some of the repairs yourself, the costs will be lower.

Exterior Doors

All exterior doors should have solid core construction and be manufactured for exterior use. Wood, fiberglass, and metal are all commonly used in exterior doors. Thump on each exterior door with your knuckles—a solid-core door will resonate with a meaty sound, while a hollow-core door will give a sound that matches its name. It's not uncommon to see an interior door used as a garage entrance, a breezeway door, or even a back door. This is not only bad for energy conservation, but also a security risk because interior passage doors offer little resistance to break-ins.

If the veneer or laminate has peeled away, or if the door is extremely warped, you're probably looking at an interior door used as an exterior unit. On metal doors, check to make sure that the vinyl or plastic trim around the window is still attached firmly—if the door's interior insulation is exposed to water or sunlight it will deteriorate. Check along the bottom of the door where sidewalk salt or accumulated dirt may have led to rusting.

Exterior doors should fit firmly in their frames without binding. Check to see that the door frames and sills are in good condition. Use your awl or fingernail to examine the sill to see if the wood is solid. If you find soft or rotted wood, make a visual note of where the door is situated, and look in the basement or crawlspace, if possible, to see if water damage has affected the joists or sill plate below.

Soft wood in the door jamb indicates water damage that may have extended to the framing. When this door was shut firmly, the wall around the door shivered suspiciously, and when the moldings were removed, rotten studs came into view.

- Cost of replacing an exterior door, slab only: $400 to $800
- Cost of combination storm window/screen door: $350
- Cost to install new locksets: $100 to $175

Windows

Windows come in a variety of styles. The most common movable windows are vertical double-hung windows that slide up and down to open and close, and casement windows, which operate by mechanical crank and generally open by pivoting sideways. Other movable windows include sliding, awning, and jalousie windows. Fixed, or nonmoving, windows include picture, bay, and garden windows, and most skylights. Each style may be single, double, or triple glazed, which refers to the layers of glass used. Double-glazed windows are more energy efficient than single, while triple-glazed units offer the best energy efficiency, especially when paired with one or more layers of "low-E" coatings on the glass.

From the outside. On southern and western exposures, windows and frames take quite a beating from the sun, wind, and precipitation. Check wooden window sills, if you can reach them, by pushing on them with your fingernail or an awl. The wood should be firm, not soft. A fresh paint job by the owners can make rotting or damaged wood look presentable from a distance, so rely on touch, not just sight.

If the window trim is vinyl or metal-clad wood, check the corners. Because both vinyl and metal expand and contract at

This cracked door jamb is the result of a forced entry. The strength of the jamb been compromised. Damage like this should raise a red flag about the safety of the neighborhood.

These cutaways show (from left to right) tempered-glass double-glazed, tinted double-glazed, double-glazed, and single-glazed windows. Double-glazed windows are more energy efficient than single-glazed windows.

different rates than wood, it is important to make sure that covering has not separated and exposed the wood below.

Check to see that no windows have cracks. In older windows, inspect the glazing to see that it is an unbroken strip all around the window pane. You may want to use your binoculars to inspect upper-story windows.

Single-glazed double-hung windows generally will have some kind of exterior storm window protection. This might mean removable wood storm windows, or combination units in either aluminum or steel. Metal combination storms should appear level and plumb. Skewed units don't offer much in the way of protection.

From the inside. Inside the house, check representative windows, if not every single one, to make sure they work. Visually inspect all double-hung windows for signs of cord breakage, or in the case of spring systems, whether the springs have been painted.

Wooden storm windows provide the same energy efficiency gain as combination storm windows, provided they're in good repair.

Combination storms have both top and bottom storm windows and a screen. Make sure no screens are missing.

Casement windows usually have snap-in internal screens. The wooden frame here is so swollen and rotten that the window is nonfunctional.

Missing glazing makes windows drafty, and can allow the wood to decay. If this problem is prevalent, make sure to check the condition of the window sashes.

Weather stripping will deteriorate over time. It's not a serious problem, but should be fixed to prevent energy loss.

Counterweight ropes in older double-hung windows wear out and break over time. The problem is annoying, but easily fixable.

(This generally decreases their operability.) Metal window guides or tracks within the window frame shouldn't be painted, because the window won't slide very well on painted tracks. Opening a double-hung window by pushing on the top rail eventually weakens the window. Look for loose joints, bowed rails, or gaps between the glass and rail. Check to see that all casement windows have cranks and screens, and operate them to see if the action is smooth.

Check windowsills to see if they are solidly fastened and free of water damage or bubbled paint. All window and door surfaces and trims should be painted or finished. Look at the wall surfaces below all windowsills and corners to check for any water damage or water staining.

In a newer home, the quality and life span of the original double- or triple-glazed windows is important. If you are looking at a 15-year-old house with double-glazed windows and notice that one is beginning to fog, it may indicate that

Deal Killers:
A fogged thermopane window has lost its seal and its energy efficiency. It no longer acts as an insulated window, and must be replaced. Several windows in this shape may indicate that all the windows in the home will soon fail. Such a situation might cost you several thousand dollars in repairs.

For garage doors, center-mounted torsion springs are preferable to extension springs that are mounted parallel to the tracks.

Newer automatic garage doors have obstruction-detecting electric eyes that prevent the door from operating if anything is in its path.

all the windows will soon fail. The fogging isn't just aesthetic; if the window is hazy, the seal is broken and the unit is no longer functioning as a double-glazed window.

In an older home with newly replaced windows, find out the manufacturer and models for the replacement windows, and, if possible, the name of the company that installed them. From this information, you may be able to find information about the window's life expectancy.

- Cost of reglazing single-pane windows: $25 to $90 per pane
- Cost of replacing entire window: $250 to $1,200
- Cost of replacing wooden storm windows with combination storm windows: $120 to $150 per window

Garage Doors and Windows

Garage doors and windows also need to be inspected. In an attached garage, the passage door between the garage and the house is considered an exterior door. It should fit snugly with weather stripping, have a dead-bolt lock, and be fire rated to resist burn-through for 20 minutes. The door between an attached garage and the house also should be self-closing to prevent garage fumes from entering the house via a door left open. Harmful exhaust vapors are heavier than air, so the garage should be at least 6 inches below the level of the living area it is attached to.

Open the large garage door and check for smooth operation.

Better garage doors are made of fiberglass or metal; older doors are sometimes made of composite wood products that can swell and become delaminated.

- Garage door replacement: 8 × 7' door, $550 to $700
- Garage door opener: $250 to $500

Interior Doors

Interior doors should swing easily and latch closed. Look for doorknob holes on the wall surface behind doors. If doors have been removed from doorways, ask if the door is in storage or if it has been thrown away. In an older home it may be difficult to find a matching replacement door. Even in a newer home, replacing doors can be expensive. Check both sides of interior doors for holes or peeling veneer. Hardwood and solid-core doors may be repairable, but hollow-core doors are difficult to fix.

- Replacement interior door, slab only: $40 to $150
- Prehung (door plus framing) interior replacement door: $80 to $200

Specialty Windows and Doors

Glass block windows in the basement or bathroom can be a real plus. These windows allow light in while providing privacy and security. Check glass blocks to see that none are broken—replacing a glass block can be a complex and expensive task.

Skylights are potential areas of difficulty if they have been improperly installed. A skylight on a roof with a less than 3-in-12, or 25%, grade must have at least a 4-inch high curb—it should not be flush-mounted. From the outside, use your binoculars to inspect the flashing and caulking around a skylight if you can. From inside, look for bubbling or water damage to the wall surfaces around the skylight. If the skylight opens, make sure that the screen is present and that whatever opening mechanism—rod, pulley system, or automatic opener—is present and works.

Check skylights from the outside by using binoculars. Look to see if metal flashings are in good shape, and if the window is mounted in a raised curb.

If You're Selling:

- All the windows should be sparkling clean. For $50 to $150 you can hire a window cleaning service to make them shine.
- All screens, including screen doors, should be free from tears. It is a fairly simple task to replace window screening.
- Clean accumulated dust from between double-hung windows and storm windows.
- Clean or paint pet-scratched exterior doors. Make sure all weather stripping is intact, or install new weather stripping.
- Oil hinges and latch mechanisms so they move freely without sticking.
- Clean and lubricate tracks of sliding doors and windows and make sure the opening hardware is firmly attached.

Cellar doors may still be found in some older homes. Check to make sure that the door is lockable and that there are holdback latches to securely hold the door when open. If it is wood, make sure that no part of the door or frame has rotted or been chewed through by animals. If it is a metal door, check the condition of the rubber weather seal.

Bifold and bypass closet doors should be checked see that both doors move smoothly on the track and close without gaps. In wide closet openings, look at the wall above the door. This wall should not sag—if it does, an inadequately sized header was used, which could be a serious structural problem.

Sliding patio doors need to move smoothly for the length of their tracks. Make sure that latch mechanisms are operable and that the frames are secure.

Patio doors are vulnerable to intruders, so look for a sturdy lock on the latching mechanism, and an auxiliary lock on the track that allows the door to open only wide enough to provide ventilation. You can also add these features yourself, of course, if you buy the house.

Check for fogging or condensation between the glass panels. Inexpensive sliding doors with metal frames often get condensation or even frost buildup in cooler weather. Check for water damage to the floors adjoining patio doors.

> • Patio door replacement: $1500 to $2000

Signs of Gold:

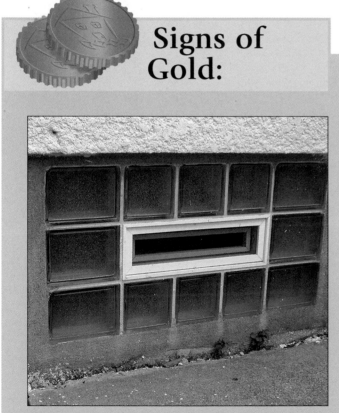

Well-installed glass block windows add security and privacy while providing light and ventilation to the basement.

Egress windows are required in all basement bedrooms. They also brighten up finished basements by admitting extra light.

In, Out, and Through: Window and Door Checklist

Item	Good Indications	Condition		
		Good	Average	Poor
Exterior				
Doors	• Firm fit, solid frame	☐	☐	☐
	• Solid core, surface in good condition	☐	☐	☐
Windows	• Solid sills and frames	☐	☐	☐
	• Glazing present, no fogging	☐	☐	☐
Garage				
Door	• Smooth movement	☐	☐	☐
	• Automatic opener	☐	☐	☐
Access door	• Solid core, exterior grade	☐	☐	☐
Interior				
Doors	• Present, good condition	☐	☐	☐
	• Door stops	☐	☐	☐
Windows	• Smooth operation	☐	☐	☐
	• Solid sills	☐	☐	☐
	• New, double glazed	☐	☐	☐
Specialty				
Patio	• Smooth operation	☐	☐	☐
	• No-lift feature	☐	☐	☐
Skylight	• Surrounding wall surface free of water stains	☐	☐	☐

Chapter 11

Indoor Weather Report
Inspecting the HVAC System

In the following few pages, I'll be talking about elements of a home that contribute to air quality and climate control. In the business, this stuff is grouped under the term *HVAC*, which stands for Heating, Ventilation, and Air Conditioning. This term generally includes the insulation as well. This means you'll be looking at the furnace, other heating appliances, fireplaces, the air-conditioning unit, any power vent fans and other venting devices, and the insulation and weather stripping.

Controlling the climate is crucial to a comfortable, healthy home, so pay close attention to these elements during your inspection. Heating and cooling costs generally are your second highest expense after your mortgage. Furnaces or central air conditioners are also quite expensive to replace, so it's important to understand the condition of these crucial units.

The heating system in any house is only as good as the insulation that keeps the outdoors separate from the indoors, so it is important to evaluate that as well.

In the interest of saving energy, super-efficient furnaces and tightly insulated houses have been the norm in new construction over the past 30 years. Unfortunately, some of these methods have diminished indoor air quality and have, in extreme cases, led to houses that rot from the inside out from the buildup of internal humidity. See Chapter 17 for a discussion of indoor air quality.

Signs of Gold: A high efficiency furnace with a multi-speed fan can save many hundreds of dollars each year in energy costs. Some such furnaces also have sophisticated multi-element filters that improve air quality enormously.

Climate control is crucial to a comfortable, healthy home, so pay close attention to these elements during your inspection.

Registers and Radiators

As you begin walking through the house, note the number and placement of registers and radiators. This will tell you much about the type of heating system found in the house.

Registers, or vents, indicate a warm-air system. A plus to look for in newer homes is a "high/low" vent system. This system takes into account the problem of using the same ducts for heating and cooling. Since warm air rises and cool air sinks, it is not ideal to have both using floor or baseboard vents. A high and low vent system means that you can open the upper vents and close the lower vents in the summer for air-conditioning, reversing them in the winter.

Radiators indicate a hot-water or steam heating system. You may see the familiar, large cast-iron radiators or smaller, baseboard convector units. The finned-tube baseboard units are more efficient and provide more even heating.

If you don't see registers or radiators in a livable room, you'll want to make a note to ask the owner why. In a newer home, there might be radiant in-floor heating, which would likely be touted on the property information form.

You may see an electric baseboard heater or a gas heater in one or more rooms—especially in new additions. Although these units do the job, electric space heaters are less efficient and more costly than heating the space via a central system, and gas-powered space heaters must be vented adequately to ensure their safety.

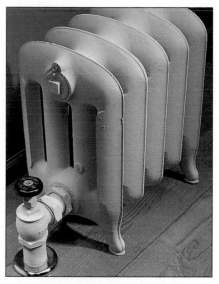

Signs of a well maintained radiator: the valves and handle are unpainted and the floor bears no evidence of water leakage.

Warm-Air Systems

Forced-air systems. The most common central heating system you will see is the warm-air furnace. A warm-air system heats air directly, and the warm air travels throughout the house via ducts. Cool air is returned to be reheated by a separate set of ducts. Most warm-air furnaces are now forced-air systems, which means a fan blows the air around.

Gravity-feed warm-air systems. Older systems rely on gravity and the physics of warm air rising and cool air sinking to circulate the air. If you walk into a basement that has very large, round ducts that gradually slope to the furnace, you've encountered the "octopus" of a gravity-feed system. This does not mean, however, that the furnace attached to those ducts is necessarily gravity fed. Some gravity-feed furnaces have been replaced with forced air without the ductwork being replaced.

Though old, these gravity-feed systems can be remarkably durable. However, you don't have the option of installing central air-conditioning with such a furnace system. A forced-air furnace, on the other hand, can be updated with a central air-conditioning unit if it doesn't already have one.

Built-in humidifiers are often poorly maintained, which can lead to mold buildup in the reservoir and potential health problems for your family. In each house I've owned, I quickly disconnected the furnace's built-in humidifier.

Examine the furnace filter. If you find a filter this dirty, it's not a large leap to wonder how well the rest of the furnace has been maintained.

Hot-Water Heat

If the house has a hot-water heating system, it will include a boiler of some sort. Even though boilers differ greatly in appearance, they all operate in the same way—water is heated, it travels to different areas of the house, and cool water is returned to be reheated.

Once again, older systems rely on gravity to move hot and cold water around the system. Newer systems incorporate a pump, which moves hot water through the system more effectively.

Many hot-water systems use oil as the heat source. Oil burners may last 30 to 50 years, but they won't be economical near the end of their age range. If the furnace or boiler is an oil burner, there will also be an oil tank. Tanks ranging up to 250 gallons can be located indoors but must be more than 7 feet from a heat source. Larger tanks are buried outdoors. An oil tank's life expectancy is about 20 years. If you need to replace an underground tank that has leaked, you will also be held responsible for removing and treating the contaminated soil, a potentially expensive and unpleasant situation.

In-floor radiant heat has multiple zones with individual thermostats and on-off valves located near a hot-water boiler.

An ancient hot water boiler can be very expensive to replace. This unit started out as a coal burner and was converted to fuel oil, then to natural gas.

Combustion Safety

Furnaces and boilers are fired by gas, oil, electricity, or possibly coal or wood. All heating units that use these combustible fuels require fresh air for combustion, and must be vented so exhaust gases are expelled outside the house. With older furnaces or boilers, metal exhaust ducts should be firmly joined with friction fittings and supported every 3 to 4 feet. This type of flue connects to a chimney—which may or may not be lined. You can look for corrosion where the pipe enters the chimney, but only a professional chimney inspection will give you information on the actual chimney condition. Because very hot flue gases heat up this vent, combustible materials (including joists) must be more than 6 inches away or protected by nonflammable barriers. Because burning fuels require an air source, enclosed furnace rooms must have adequate ventilation via grilles or louvered doors. In newer furnaces, combustion air may be piped in directly from the outdoors. An electric warm-air or hot-water heating system does not produce combustion gases or flames, so it can be located anywhere in the house.

The new, super-efficient furnaces remove so much heat from the fuel burning that the exhaust gases can be vented through a PVC pipe. You will see two such pipes leading in from the side of the house, or perhaps from the roof. One is an intake pipe and the other the exhaust pipe. When you inspect these vents, check to see that they are not covered or blocked by vegetation or structures. The high-efficiency furnaces are condensing furnaces, so you also need to look at how the condensed moisture is drained. There may be a hose directly from the furnace to a floor drain, or a small pump may be located near the furnace. The pump allows the hose to be installed across the ceiling and out of the way.

Electric Baseboard Heaters

The colder the climate in your area, the less likely it is that you will see an entire house heated by electric baseboard heaters because they are fairly expensive to run. Look for units that have a thermostat with an "off" position. If there is no off selection for the thermostat, the heater must be turned off at the breaker box. Electrical cord insulation can be melted by the heat of electric baseboard heaters, so make sure there are no electrical receptacles located above the heater.

A rusty furnace flue may leak carbon monoxide into the house. This is a dangerous situation that needs immediate attention.

This older gas-fired furnace should not be enclosed in this closet, because furnaces require plenty of ventilation. This installation is a code violation. Without louvers in the closet doors and grates built into the walls, this furnace has no adequate source of combustion air.

ELECTRIC BASEBOARD HEATERS

- Life expectancy: 15 years
- Replacement cost : $100 to $200 each

Photo courtesy of Weil-McLain

Baseboard hot water radiators are the sign of a newer hot water system.

A heat pump system resembles a central air-conditioning unit, both in appearance and function.

Heat Pumps

In milder climates, heat pumps are used for both heating and cooling. A heat pump works just like an air-conditioning unit or refrigerator. In cool weather, it pulls heat from outside air and moves it inside. In hot weather, it pulls heat from the inside air and moves it outside. A heat pump system may be "split"—one half inside and one half outside, or "packaged"—all-in-one outdoors. Neither type should be located in the direct sun or under trees or shrubs, and neither should be subject to roof run-off. The coils, cooling fins, and casing should be clean and in good condition.

Another type of heat-pump system is the geothermal heat pump. This system is based on the fact that ambient ground temperature is about 55°. Water is pumped through a series of underground pipes to either cool off or warm up. The heat pump either extracts heat or dissipates it using the water. There are two geothermal system types—closed or open. A closed system continually recirculates the same water. An open system continually pulls in water from a pond or river, runs it through the system, and dumps it. In some locations open systems may no longer be legal. Geothermal heat pumps are located inside the house.

Fireplaces and Woodstoves

Fireplaces and woodstoves have a romantic appeal, but it's rarely appropriate for them to be the sole heat source. All fireplaces should have a damper, which opens and closes the flue. There should be a handle or chain inside the fireplace that operates the damper. Test it, even if it is sooty. Ask how often the fireplace is used and when the chimney was last swept and inspected. A chimney built before 1950 is probably not lined, which increases the chance of fire or carbon-monoxide poisoning. Note that glass doors and screens should remain with the house, but fireplace tools generally do not.

Woodstoves are more efficient than fireplaces because they have dampers to control air flow, and heat radiates from all sides of the stove. The proper installation of a woodstove and chimney is important for safety. The stove must be situated on a nonflammable surface and have 36 inches of clearance between it and any flammable materials. The clearance between the flue pipe and combustibles must be a minimum of three times the diameter of the flue pipe.

Air-Conditioning

In your perusal of the home's exterior, you will have noticed if there is an air conditioner condenser or heat pump unit. This unit should be located past the roof drip line and should have a minimum of one foot of clear space on all sides. The cooling fins should be intact, not bent or corroded. You may see window air conditioners, but these will not necessarily stay with the property. If you see both, you will want to know why—a heat pump or central air conditioner should serve the whole house.

A newer system may be up to three times more energy efficient than a unit that's even 10 years old. With central air conditioners, bigger is not necessarily better—an oversized system cycles quickly and may not dehumidify the air properly. A unit that's too small will dehumidify the air, but won't cool it sufficiently. Cooling capacity is measured in tons or BTUs. Figure 1 ton or 12,000 BTUs per 550 square feet to be cooled. A central air conditioner is more efficient than window units if you are planning on cooling the whole house.

AIR CONDITIONING

- Central air-conditioning units
  Life expectancy: 10 to 15 years
  Replacement cost: $1,000 to $3,000
- Window air-conditioning units
  Life expectancy: 10 to 20 years
  Replacement cost: $400 to $1,000

Insulation

Insulation is an important part of the heating and cooling system, but you may not be able to see any of it. From outside the house, you may see round patches in the exterior siding. This is a sign that blown-in insulation has been added. Inside the house you will want to look in the attic and in crawl spaces. It is a good idea to ask before opening ceiling attic access panels—if the

Freestanding heating stoves require fire-resistant surrounds to prevent damage from heat. This ceramic tile surround forms an ideal, safe setting for a gas or woodstove.

Cracks in fireplace mortar can allow hot gases to enter wall cavities in the home. The mortar must be repaired before the fireplace is used.

This brand-new AC unit sits on a cement pad that is still curing. You can plan on 10 to 15 years of good service from such a unit.

Corroded cooling fins on this AC unit were caused by a family dog urinating on the unit. The unit's efficiency has been compromised, and it should be replaced.

house has blown-in insulation you might get a face full of cellulose when you push up a panel. Ideally, such a panel should be caulked between the panel and the frame to prevent air and moisture flow.

In a house with an unfinished attic and attic access, you can examine the existing insulation. The minimum recommended ceiling insulation depth is 12 inches. In order to achieve this, most attics have insulation between the joists with additional insulation on top of the joists. The additional layers should be laid perpendicular to the joists. If possible, you want to see if the insulation is blocking air flow by covering soffit vents. Air needs to circulate under the roof to prevent ice dams in cold climates and buildup of excessive heat in the summer. Finished attics and cathedral ceilings are difficult to evaluate.

Attic insulation should be dry and should not have items stored on top of it. Wet insulation in the attic is a sign of a leaky roof or extensive water vapor escaping from the house. Insulation that has gotten wet may or may not retain its insulating properties after it dries.

Floors above crawl spaces should be insulated, but this is a difficult and dirty inspection job. If there is a crawl space present, it should have ventilation to the outside every 6 to 8 feet. If not, it is likely that there is moisture buildup. If you can see into a crawl space, take a look. The insulation should be secured under the joists (not hanging down) with the vapor barrier against the floor, not facing outward.

Weather stripping is important for stopping drafts and preventing dirt and insects from entering the home. Though you'd prefer to find it already installed, weather stripping is an inexpensive and easy do-it-yourself project.

This attic insulation leaves something to be desired in the installation, but it features an 8″ layer of fiberglass over 10″ floor joist cavities filled with blown-in cellulose—an acceptable amount of insulation, even for very cold climates.

Ventilation Systems

While older homes have many air leaks that provide some fresh air exchange, newer homes are built with materials and processes that can leave them virtually airtight and susceptible to air quality problems. The newer, more energy efficient, and better insulated a home is, the more important proper ventilation is. There are a number of different ventilation possibilities. The drafts of an older home are uncontrolled ventilation—these homes aren't very energy efficient, but they rarely have serious air quality problems. Spot ventilation in the form of bathroom fans and kitchen exhaust hoods are familiar to most of us. This is controlled ventilation. Newer, more airtight homes need to have automatic controlled ventilation. Automated ventilation removes humidity, carbon dioxide, and other pollutants from the home and brings in outside fresh air. An automated ventilation system may or may not have a heat exchanger. If it doesn't, heated or cooled air will be wasted when it is vented. An automated ventilation system with a heat exchanger may be located in the attic or the basement. It is likely to be integrated with the heating and cooling system, but it may not be.

Kitchen and bathroom exhaust fans must vent to the outdoors. Some kitchen fans are recirculating air filters—this is not sufficient. Bathroom fans are sometimes vented through the ceiling into the attic, which violates building codes.

A turbine vent may actually draw warm air out of the house in the winter if it can't be closed off. A home with several turbine vents is probably over-ventilated.

Signs of Gold:

This air-to-air exchanger is an excellent ventilation system. It swaps stale indoor air for fresh outside air without losing heat in the winter or cool air in the summer. A home with this kind of system will have excellent air quality, which is especially important if any members of your family suffer from asthma or other respiratory problems.

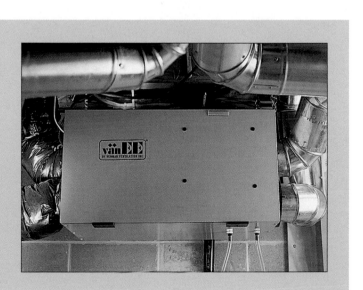

Indoor Weather Report: HVAC Checklist

Item	Good Indications	Condition		
		Good	Average	Poor
Furnace	• High efficiency	☐	☐	☐
	• Less than 10 yrs. old	☐	☐	☐
	• No asbestos on ducts or pipes	☐	☐	☐
	• Proper combustion ventilation	☐	☐	☐
Radiators	• Valves not painted	☐	☐	☐
	• No signs of water leakage	☐	☐	☐
Vents	• Dampers on ducts/ louvers funtional	☐	☐	☐
Air conditioner or heat pump	• Fins and coils clean	☐	☐	☐
	• Away from roof drip line	☐	☐	☐
	• Less than 10 yrs. old	☐	☐	☐
Fireplace & Woodstove	• Lined chimney	☐	☐	☐
	• Damper works	☐	☐	☐
	• 36" clearance around stove	☐	☐	☐
	• Chimney cleaned recently	☐	☐	☐
Ventilation	• Kitchen hood vented outside	☐	☐	☐
	• Bathroom fan vented outside	☐	☐	☐
	• Heat exchanger	☐	☐	☐
Insulation	• Sufficient depth	☐	☐	☐
	• Dry, uncompressed	☐	☐	☐
	• Crawl space insulated	☐	☐	☐

Anatomy of HVAC Systems

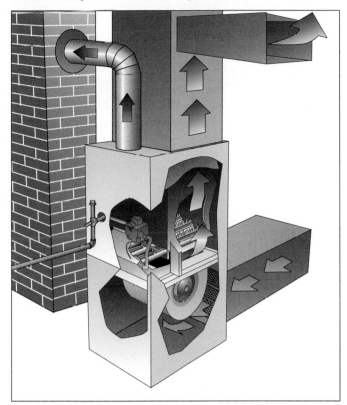

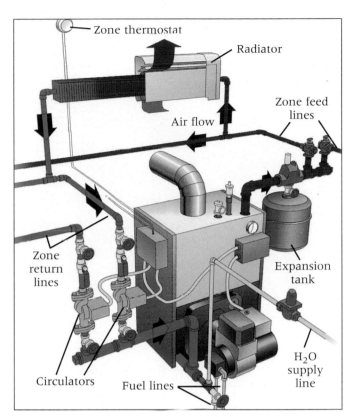

Warm-air systems all work in much the same way. Cold air is drawn into the bottom of the furnace, moves around the heat exchanger, where it is heated by burning fuel, and is forced through the top of the furnace into ductwork by an electric fan. Combustion gases from the fire chamber are vented into a chimney or exhaust flue.

Hot water systems circulate water through a closed system of pipes that pass through a boiler. Newer boilers feature one or more circulator pumps to move water through the system, while older boilers work through simple convection. A system of radiators transfers the heat from the water to the air in the various living spaces.

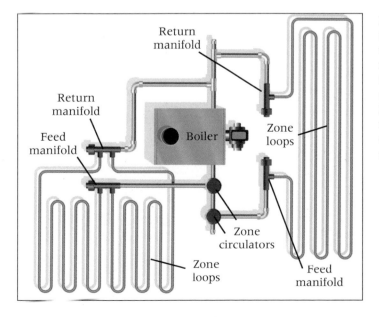

Radiant floor systems use circulators to move hot water from a boiler, and manifolds to divide the system into zones that can be controlled separately. Radiant heat systems are as much as 30% more cost-efficient than other heating systems.

Chapter 12

Watts It All About

The Electrical System

Electrical mishaps are one of the few home problems that can have instant, devastating consequences.

Even professional home inspectors quite often defer to an electrician when it comes to wiring.

Inspecting the electrical system makes almost everybody a little nervous, and this is one instance where the concern is warranted. A bad electrical system at best is expensive to upgrade or replace; the worst-case scenario can be catastrophic. Faulty wiring is a leading cause of home fires. Even professional home inspectors often defer to an electrician when it comes to wiring. So don't assume that the information in this chapter is all you'll need to judge the quality of the wiring system. What I can do, though, is point out some clear warning signs that further inspection is mandatory. For the most part, these are all visual clues that won't require you to touch any wires or electrical devices.

There's no reason to be uneasy when inspecting wiring, because you won't need to do anything but look, operate switches, and test outlets. Pay close attention to what you see, but under no circumstances should you remove the covers on outlets, switches, electrical boxes, or the main service boxes. Only a qualified electrician should be making this kind of inspection.

TYPICAL ELECTRICAL REPAIR COSTS	
Item	Average cost
Update knob & tube service	$4,000 or more
Increase from 30 to 100 amps	$800 or more
Increase to 200 amps	$1,400 or more
Replace service mast	$500 or more
Install new 120-volt circuit	$250 or more
Install dryer circuit	$175 or more
Replace fixture/switch/receptacle	$50 or more

The Service Entrance

Begin your inspection outside the house—where the wires enter. In newer homes or in neighborhoods that have had major utility upgrades, the service wires enter underground. If this is the case, just skip to the next section, where I discuss the service panel. But if the service wires enter the home through overhead wires, begin your inspection by looking at these wires and the pole where the wires are attached to the house, called the *service mast*.

You'll learn something important just by looking at the number of wires that enter the mast. If there are only two wires, it means that house is served by 120-volt current, which will make it impossible to use 240-volt appliances—a stove, clothes dryer, or water heater, for example. It also means that house probably is served by only 30 amps of power—woefully inadequate for today's home.

Look at the condition of the mast. The mast should be securely anchored, and the wires solidly attached to it. The wires should be arranged in a "drip loop" that will allow rainwater to drop straight down rather than run into the service mast. Also look at the wires running from the service mast to the utility pole. Make sure the wires are well above any walkways and roof structures. Most electrical codes require that service wires be at least 10 feet above walkways and steps, 12 feet above driveways, and 3 feet above roof surfaces.

Follow the service mast down the side of the house, and make sure it is firmly attached all the way to the electric meter, where the service enters the house. Look at the metal box that contains the meter, and make sure it is weather tight and has no signs of rust or indications of water damage.

The Service Panel

Now go indoors and look for the main service panel, which is usually located in the garage, basement, or another utility area. The service panel is often a gray metal cabinet. Provided the floor around the panel is dry, you can—and should—open the access door on the panel. But under no circumstances should you unscrew and remove the cover on the panel.

If there are two or more service panels, one is usually the main panel, and the others are subpanels added at some point because the main panel had no more room for additional circuits.

In most situations, the main service panel contains circuit breakers that control and protect the individual circuits that run

The service mast should be well anchored, and shaped with a drip loop that prevents rainwater from entering the mast.

The electric meter should be in good condition, with no signs of rust or water entry.

Signs of Gold:

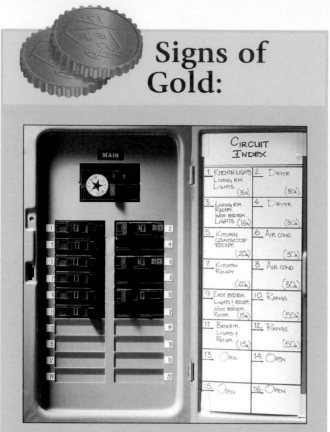

An amply sized service panel with open circuit breaker slots for expansion, and a clearly legible index, are signs that the electrical service has been competently installed.

A combination circuit breaker panel and fuse box is common. There's nothing wrong with this scenario, provided the service is adequate in size.

through the home. Have a look at the main circuit breaker, which is usually located at the top or right of the panel, centered between two columns of individual circuit breakers. Read the number on the main breaker—this will tell you the amperage of the service. A bare minimum for a small house is 60 amps, while 100 amps is much better. For a medium to large house, 100 amps might be suitable, but 150 or even 200 amps is better.

It's a very good sign if you see a neatly organized panel with a clearly marked index identifying the circuits, and several open slots in the panel, which will allow you to add additional circuits. I've even seen panels where the individual cables running from the box are clearly labeled to indicate which circuits they serve. This kind of meticulousness tells you that the service was professionally installed and is unlikely to cause you problems.

I look with extreme suspicion on any service panels that are protected by fuses instead of circuit breakers. These are virtually always outdated and will likely require upgrading very soon. It's a bit less troublesome if it's the subpanels that contain fuses, provided the main service panel contains circuit breakers. But by and large, the presence of any fuse panel indicates a service that you'll want to have inspected by a professional.

The electrical code requires that all main service panels be connected to ground through a bare copper wire that runs to a grounding rod imbedded in the ground, or, in older installations, to a metal water pipe. If you don't see this, make a note of it.

Circuits

Here are some of the most common and important circuit requirements from the National Electric Code. If a house you're interested in doesn't comply with each and every one of these, it doesn't necessarily mean there is a major problem. Many of these requirements are recent

additions to the code, which means that most older homes will not comply with all of them. But it will be a very good sign if you see that the home does meet most or all of these stipulations. Reading the index at the service panel should tell you all you need to know about the circuits that run through the house.

- Bathrooms should be serviced by a dedicated 20-amp circuit—not a general lighting circuit that serves other areas of the home.
- Kitchens should have two 20-amp small appliance circuits that feed receptacles, plus 15-amp dedicated circuits for the dishwasher and garbage disposer. In addition, there should be another circuit that controls general lighting.
- Laundries should have a separate 20-amp circuit for the washing machine. The dryer, if electric, should be fed by a 30-amp, 120/240 volt circuit.
- Living rooms, bedrooms and other general living spaces should be served by at least one 15-amp general lighting/outlet circuit for each 600 square feet of finished living space. Living rooms should be served by at least two circuits.
- Hot tubs, whirlpools, and other water features should be served by GFCI-protected circuits.

Many of the requirements of the National Electrical Code are recent additions, which means that most older homes will not comply with all of them.

SNO-FLEX E-7477-S 12AL/2 WITH GROUND TYPE NM 600V

Look at the sheathing on electrical cables to determine the type of wiring. If the letters on the cable sheathing indicate the circuits are wired with aluminum wiring, have an electrician inspect them. Aluminum wiring can pose severe a fire hazard, unless it has been installed and updated by professionals who specialize in aluminum wiring. The details are too complicated to address here, but if you

The letters AL stamped on cable sheathing indicate aluminum wiring. Have a qualified electrician inspect any suspected aluminum wiring.

Surface-mounted porcelain insulators indicate very old, knob-and-tube wiring. If this is still active, you're looking at an antiquated system that will need to be updated at major expense.

Deal Killers:

Very old wiring systems may use wire routed in a "knob-and-tube" system, where the individual conductors are surface mounted and held by ceramic insulators. You'll see the clearest evidence of this in basements and other unfinished spaces. If this is the active wiring system, you're looking at a very antiquated system that will almost certainly require major and expensive rennovation. But it's also quite common that the knob and tube system is simply a remnant from the original system, and is no longer active at all. Make sure to check this out. If the active system is truly a knob and tube arrangement, make sure to factor in the cost of a major upgrade before you make an offer on the house.

determine that the house has aluminum wiring, always have a secondary inspection by an electrician with experience in that area.

Follow the circuit wires from the service panel. If the space is unfinished, the cables should be securely stapled to the sides of framing members—or should be routed through holes drilled in the framing members. At junction boxes—usually metal boxes where circuit wires branch or meet—there should be secure covers, and the cables should be clamped in place.

Outlets and Light Fixtures

As you go through the home room by room, look at the number and condition of the electrical outlets, switches, and light fixtures. Be aware that it's very, very common for older homes to be slightly below par when it comes to complying with the latest electrical code requirements. Don't fret about this too much; it's not a reason to reject an otherwise desirable home.

Ideally, most living spaces should have receptacle outlets spaced no more than 12 feet apart, so that no point is more than 6 feet from a plug-in. In kitchens, receptacles should be spaced no more than 48 inches apart behind the countertops. In kitchens, bathrooms, basements, and anywhere there is the possibility of contact with water, the outlets should have GFCI receptacles for safety.

Look at each receptacle to see if it has a grounding opening that will allow it to accept three-prong, grounded plugs. It's not

Cracked or broken outlets are not expensive to fix, but they pose a fire danger if they're not addressed.

uncommon or dangerous for older homes to still have two-slot outlets, but it would be a good idea to convert them to new receptacles at some point. Use a plug-in circuit tester to check the wiring on each outlet. A combination of neon lights on the test can tell you if the outlet is properly grounded, or if the circuit wires have been reversed, which is a potentially dangerous situation. One or two outlets that show problems is no big deal. But if you discover that many of the outlets are not grounded, the entire electrical system might be faulty—a sign that you'd want to have an electrician take a much closer look.

Tracks of the Poor Craftsman:

Substandard repairs are a concern in the electrical arena. Anytime I see signs of shoddy work on the wiring, I dig deeper to make sure things are safe. Here are some signs that the wiring system may have been adapted or installed by a poor craftsman:

- Wall switches installed upside down, so the "ON" markings read upside down.
- Outlets or switches at unusual heights on the wall. Pros install outlets at about 12" above the floor; switches at 48". In kitchens, pros install outlets roughly at the midpoint of 18" backsplashes above countertops.
- Electrical boxes installed so the front edge is not flush with the wall surface.
- Electrical cables that aren't securely stapled or anchored to framing members.
- Electrical connections made outside an electrical box.
- Track wiring for receptacles and light fixtures. This system is an easy way to extend circuits without routing cables through walls, and is often installed to cut corners.

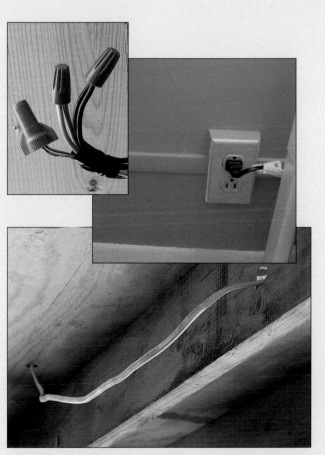

Common indications of poor wiring, from top: wire connections made outside an approved box; outlets installed with track wiring products; exposed cables not adequately anchored to framing members.

The electrical code requires that each living space have some kind of light fixture controlled by a wall switch. This means that either there should be a ceiling fixture controlled by a wall switch or a wall switch-controlled wall outlet where a floor or table lamp could be plugged in. But again, it's pretty common in older homes to find a room or two that doesn't meet this requirement.

Watts It All About: The Electrical System

Item	Good Indications	Condition		
		Good	Average	Poor
Service mast & entrance	• Securely anchored	☐	☐	☐
	• No rust, weather damage	☐	☐	☐
Service size	• 100 amps for small house	☐	☐	☐
	• 150 amp minimum for larger house	☐	☐	☐
Service panel	• Breakers, not fuses	☐	☐	☐
	• Empty slots for new circuits	☐	☐	☐
Circuits	• Dedicated appliance circuits	☐	☐	☐
	• Kitchen, bath circuits	☐	☐	☐
Receptacles	• GFCIs in kitchen, bath, garage, etc.	☐	☐	☐
	• Spaced conveniently in living areas	☐	☐	☐
	• Grounded, in good condition	☐	☐	☐
Light fixtures	• Switchable lights in every living space	☐	☐	☐

Anatomy of an Electrical System

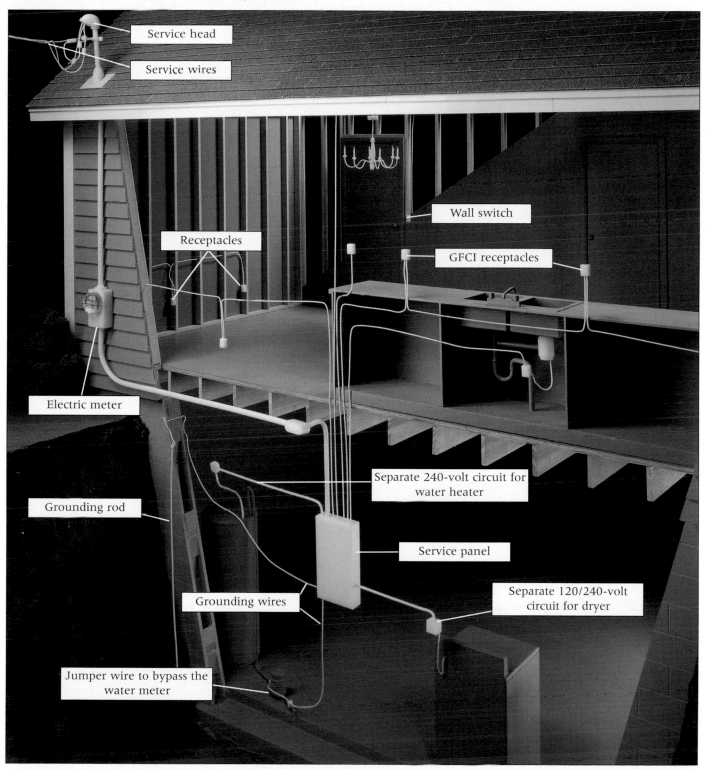

Service head

Service wires

Wall switch

Receptacles

GFCI receptacles

Electric meter

Grounding rod

Separate 240-volt circuit for water heater

Service panel

Grounding wires

Separate 120/240-volt circuit for dryer

Jumper wire to bypass the water meter

A typical electrical system routes through a home as shown in this model. The key points to inspect: the service mast and electric meter, the service panel, the circuits, the receptacles and switches.

Chapter 13

Plumbing the Depths

Assessing the Plumbing System

An all-copper water supply system is a positive sign, provided it has been correctly installed.

Earlier in this book, I've made the point that water is a cruel villain when it infiltrates from the outside in the form of rain and snow. But water is also an ally to a comfortable, healthy life, and in this chapter we'll look at the plumbing system that brings in fresh water and removes wastes and carries it to a sewer or septic system. Although the plumbing system is not quite as dramatic or potentially dangerous as the wiring system, serious problems with it can be equally expensive. An up-to-date plumbing system that has been well installed and well maintained is one of the most important virtues to look for in a home you are considering.

Although the plumbing system is not quite as dramatic or potentially dangerous as the wiring system, it carries equal capacity for expensive problems.

ITEM	LIFE EXPECTANCY	AVERAGE COST OF REPLACEMENT
Fix leaky faucet	2 to 4 years	$50 and up
Replace faucet	7 to 10 years	$100 and up
Replace sink, kitchen	20 years	$350 and up
Replace toilet	20 years	$350 and up
Install new fiberglass tub/shower	30 years	$750 to $3,500
Replace cast iron tub	40 years	$975 and up
Replace ceramic tile shower pan	20 years	$1,250 and up
Replace septic tank	10 years	$5,000 and up
Frame, install new bathroom (three-piece)		$7,500 and up

Like the electrical system, a plumbing system can be very complicated, so it makes good sense to have a licensed plumber inspect the system if you have any doubts about it whatsoever. On the following pages, I'll show some of the warning signs that indicate you're looking at a plumbing system in trouble.

The Water Supply System

Start your inspection in the basement or utility space where the water service begins its route through the home. This is usually found near the water meter. The water supply pipes are between ½" and 1" in diameter, and run in pairs of parallel hot- and cold-water pipes through most of the house. At the entry point, the pipe should be at least ¾", and preferably 1" in diameter. If the house has ½" diameter pipe entering the home, the water pressure is likely to be insufficient to adequately supply all the fixtures. Make sure there is a main shutoff valve. Usually, though not always, this is found near the water meter.

Closely examine the pipes to see what materials have been used. Be aware that a typical house can have several hundred feet of supply pipes, and that several materials may have been used, since homes are often upgraded in piecemeal fashion. Upgrading an entire water supply system can easily cost $10,000 or more, so this network warrants careful inspection.

Copper is generally regarded as the best material for plumbing pipes. Copper pipes have a distinctive orange color; if painted, you may see green oxidation bleeding through the paint. In addition, painted copper pipes can be distinguished from steel pipes because a magnet won't be attracted to copper.

Plastic pipes come in several forms, but none is regarded as a very good material for supply pipes. PVC plastic, usually white, and gray polybutylene are now regarded as dangerously substandard materials for water supply pipes; if you find these, be prepared to shell out for an upgrade to copper. Cream-colored CPVC plastic is still allowed as a water supply material in some code areas, but it, too, is generally regarded as substandard.

Galvanized steel was once the norm but is now rarely used in new construction. When used at all, it's installed to repair damaged pipes in an existing galvanized steel system. If you find this material used predominantly in the plumbing system, you're looking at a pretty old system that will require updating at some point. Carefully inspect the joints of these pipes, which is where problems are first evident.

Water supply pipes you might find include, from top: galvanized iron, CPVC plastic, PB plastic, rigid copper, chromed copper supply pipes, and flexible copper.

WATER SUPPLY PIPE MATERIALS

MATERIAL	LIFE EXPECTANCY
Galvanized steel	40 years
Plastic	10 years
Copper	50 years and up

Converting galvanized steel to copper

ACCESSIBILITY	INSTALLATION COST
Exposed	$11 per foot
Hidden in wall	$35 per foot

Deal Killers:

Rust stains in a toilet, tub, or ceramic sink may indicate that iron pipes are corroding and may be on their last legs. There can be other causes for this symptom—heavy mineral concentrations in the fresh water supply, for example—but when you see this kind of rust staining, be on guard.

Rust stains in sinks and tubs may be a symptom of iron water supply pipes that are seriously corroded. Have a licensed plumber inspect the situation.

Improper water supply transitions have dissimilar metals—in this case copper and galvanized iron—in direct contact with one another. Proper installations use dielectric unions (below), which isolate dissimilar metals and prevent them from touching one another.

Transition fittings. In many homes, different supply pipe materials will be joined together. Be on the lookout, especially, for copper pipe joined to galvanized iron. If these materials are joined directly together, an 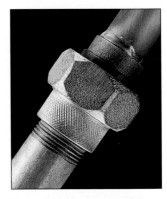 electrochemical reaction will gradually cause the joints to break down. The proper way to make such a connection is with a dielectric union.

Tracks of the Poor Craftsman:

Plastic water supply pipes in any form are a strong indication of suspect craftsmanship. Few reputable professional plumbers or savvy do-it-yourselfers install plastic water supply pipes.

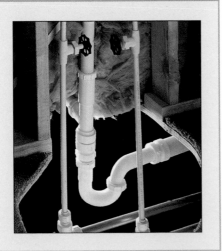

DWV (Drain-Waste-Vent) System

Unike water supply pipes, where the presence of plastic is a warning sign, plastic drain pipes are a good sign, provided they are installed correctly.

DWV pipes will generally be made from white PVC plastic or black ABS plastic. The only warning sign to watch for is PVC and ABS joined together. These plastics shrink and expand at different rates, so they should never be joined together.

It's generally fine, though, for plastic pipes to be joined to metals in the DWV system. This is sometimes done with threaded fittings or, in larger pipes, with rubber sleeves and clamps. Inspect these joints carefully, though, to make sure they're not damaged.

ABS plastic is black and was the first type of plastic used in home drain systems. It is very durable, but some codes now restrict its use in new installations, especially where it is exposed to sunlight.

PVC plastic is white, and is generally regarded as the best material for drain systems. If you see a drain system made entirely of PVC plastic, this is a good sign—the installation is likely to be relatively new.

Cast iron and galvanized iron are older materials. Cast iron is used for main vent stacks; galvanized iron for smaller branch drains. Cast iron is very durable, but galvanized iron tends to corrode, so you should inspect it carefully, especially at joints.

Copper and brass are also sometimes found in drain lines, although this is somewhat rare. It's generally a good sign, though: these metals age very slowly and provide many years of good service.

Lead is a soft metal that is occasionally found in very old plumbing systems. You may even be able to dimple it by squeezing it with your hands. Lead pipes are not a health hazard when used in drain lines, but they do indicate an old system that probably is in need of upgrading.

In the utility space, look to make sure there is a main cleanout opening for the sewer line. There should be a cleanout fitting on the main waste-vent stack, and another located somewhere along the horizontal sewer line leading out to the street. Without these fittings, it will be very difficult to clear major clogs, should they ever occur.

When you talk to the owner, make sure to ask about routine service or any major repairs on the sewer system. It's a good sign if you hear that the sewer lines to the street have been periodically scoured out to keep them flowing freely.

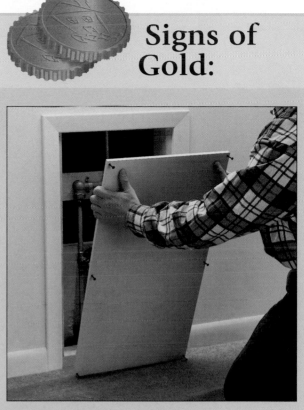

Signs of Gold:

Look for access panels that give exposure to the connections for all plumbing fixtures in a bathroom. An access panel will make future repairs much easier and less expensive.

When you talk to the owner, make sure to ask about routine service or any major repairs on the sewer system. It's a good sign if you hear that the sewer lines to the street have been periodically scoured out to keep them flowing freely.

Signs of Gold:

This drain pipe has an elbow of PVC plastic used to joint two lengths of black ABS pipe. This is a sign of poor craftsmanship. Either material is acceptable; mixing them is a major mistake.

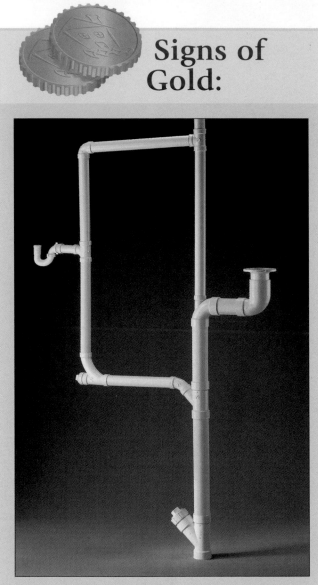

A drain system made entirely of PVC plastic is a new system that will give years of trouble-free service.

Brass and copper drain pipes, when installed in a newer house, are always a sign of a homeowner willing to invest in the very best, since these materials are much more expensive than the plastics that are more commonly used.

The Vent System

In rare instances, you may find a house with a drain system that is inadequately vented. Vent pipes are the means by which fresh, outside air is introduced into the system. Without this, flowing water in the drain system can create suction that empties drain traps and allows sewer gases to rise into the home. If you notice a smell of sewer gas anywhere in the home, or if you hear a distinct sound of gurgling in the drains when you run water in a sink or tub, it's possible the drain system isn't vented correctly. This can pose a health hazard, so regard this seriously if you find it.

Fresh air enters the system through vent stacks, usually on the roof of the house. From outside the home, look for short chimney-like stacks penetrating through the roof. Ideally, there should be one vent for each main waste-vent stack in the home. In a small home, there should be at least one such stack; most homes will have two or more. Often, these can be seen extending above the bathrooms or the kitchen. If you don't spot a vent stack, it's possible that the system is vented into the attic. This is a hazardous installation that should be corrected before you move into the house.

Look for a main cleanout fitting somewhere on the soil stack or along the main drain. Access to the cleanout is crucial in the event that the main drain becomes clogged.

Utility Areas

- Are exposed pipes properly supported with pipe straps made from the same materials? Plastic pipes should be supported with plastic straps; copper with copper, and so forth.
- Where shower and toilet drains are visible from basement or utility spaces, look for signs of water damage. Mere discoloration may indicate simply that there was a leak in the past. Probe these areas with a screwdriver. If you encounter soft wood, it's an indication that there is active rot occuring.
- Look for water supply pipes leading to outdoor hose bibs. Does each such line have a shutoff valve that allows the spigot to be turned off in winter?
- At the water heater, look to see if the water pipes are connected with proper fittings
- Make sure the water heater has a T & P (temperature and pressure relief) valve and discharge pipe. Flip the lever briefly to make sure it allows water to vent.
- Though they were popular at one time as an energy conservation measure, insulating blankets around a water heater are a bad idea—they also may void the manufacturer's warranty.

The short "chimneys" you see on the roof are vents for the plumbing system. If you don't see them, it means the system may be improperly vented—possibly into the attic.

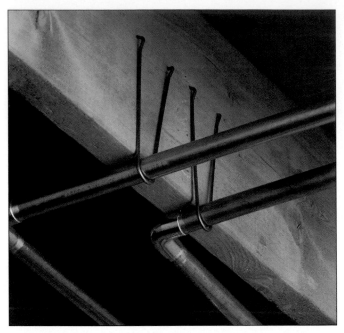

Plumbing pipes should be properly supported with straps made from the same materials as the plumbing pipes.

- On gas water heaters, make sure the exhaust flue runs at a slight upward angle to the chimney. To test the draft of the flue, turn the water temperature setting to "high," then place a lighted match near the front of the flue; you should see the flame deflect toward the flue, or even blow out entirely. If not, it means the chimney is not drawing exhaust gases properly, and that carbon monoxide could be backing up into the home.

Kitchen

Open the doors of the sink cabinet and look for signs of moisture damage—blistered paint on the wall, warped wood on the floor of the cabinet. Make sure there are shutoff valves for both the hot and cold faucet lines and for the supply pipe that feeds the dishwasher. If these aren't present, you'll want to install them soon after moving into the house. It's a bad sign if you

If the subfloor around this shower drain is soft when you probe it with a screwdriver, there is serious rot afoot. If the wood is dry and solid, though, the water infiltration was brief and did not cause major problems.

Look for shutoff valves on pipes running to exterior hose bibs.

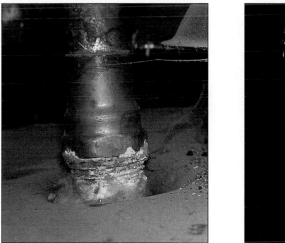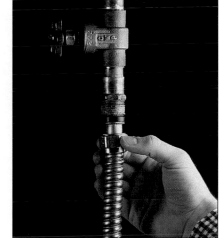

Inspect the water pipe connections to the water heater. Copper attached directly to threaded iron fittings (left) can create serious corrosion. Use of heat-saver nipples (right) is the more correct way to make these connections.

Test the T & P valve on the water heater by flipping the lever. A missing or inoperable temperature and pressure relief valve is a code violation that needs to be addressed.

see bottles of partially used liquid drain cleaner stored below the sink; not only does it indicate chronic backup problems, but possible damage to the pipes, since these chemicals can be destructive.

Examine the food disposal—does it have a reverse/reset switch that can be reset if the unit jams? Can you read a manufacturing date on the disposal, which will tell you its age? (Ten years is a typical life span.)

Test the faucet by turning it on and off several times, looking for leaks around the base of the faucet. Repairing such a faucet is merely an inconvenience that would never affect your decision to buy the house, but the fewer such inconveniences, the better. More to the point, though, if there is seeping water around the base of the sink, it may have infiltrated the substrate on the countertop, causing rot or softening.

Run a full sink of water with the drain stopper in place, then remove the stopper and listen and watch as the water drains. Look at the drain trap and drain arm to see if moisture beads up around the joints, and listen for the sound of gurgling. This is especially likely in sink island installations, where the drains may not be properly vented.

You're entitled to run the dishwasher and other appliances—if not on this visit, then certainly on a follow-up. If they're included

Take a long hard look under the kitchen sink, looking for signs of ongoing water damge, and evaluating the food disposal and other appliance hookups.

This shower faucet has a huge gap around the handle that allows water into the wall. Close inspection revealed severe decay in the framing members within this wall.

Failed grout joints in a shower wall allow enormous damage as water seeps back behind the tile. The signs can be so subtle that entire walls can be destroyed before you even have a clue that the damage is happening.

in the asking price of the home, you've a perfect right to determine their condition. If possible, remove the bottom panel on the dishwasher and look for any signs of water damage, indications that the drain hose or water supply line is leaking.

Likewise, examine the floor around the refrigerator to make sure the ice maker supply tube is not leaking.

Bathrooms

Here, you'll be checking the operation of the sinks and toilets, obviously. But more critical problems will be found around the tub and shower, where splashing water can cause hidden damage to wall and floor structures. Major problems can originate with deceptively mild, almost invisible signs, so take care here.

As you did in the kitchen, test the faucets at the vanity and tub and shower. Make sure there are shutoff valves that can be closed to shut off the water when repairs are necessary. On tubs and showers, these valves should be found behind access panels.

While inspecting inside the access panel, look to see if the back side of the wall surfaces are visible. If this is the back side of any ceramic tile installation, the ideal backing surface is cementboard, which is gray and has a rough texture. If you find this, it also means that the tile job is probably no more than 10 years old, since cement board is a relatively new building material. Or, if the tile installation is slightly older, you might find water-resistant wallboard—sometimes called green board or blue board. This, too, can be decent backing material for ceramic tile. But if you find that ordinary wallboard has been used as the backer for ceramic tile, it's likely the wall will eventually fail and you'll have to redo it.

Pay particular attention to the joints around ceramic tile tubs and showers. Even very small cracks can allow a surprisingly large amount of water into a wall—one plumber told me that service calls to repair "leaky pipes" are, in about 75% of all cases, originated by ceramic tile installations with compromised joints.

Look very closely at every ceramic tile installation. Press lightly on the walls with your hands. If water has infiltrated the wall, you may feel the wall "give" under light pressure. This can be a very serious problem that leads to repairs costing thousands of dollars. This is especially likely on ceramic tile walls installed over ordinary wallboard.

In a shower with a ceramic tile base, run water and look to see if the water puddles rather than running down the drain. You may see signs that the shower pan has buckled or warped, an indication that the shower will need to be rebuilt shortly.

Examine the caulking used to seal the joints between the tub or shower and the walls and floors. If you see a very thick bead of brand-new caulk, you can be suspicious that the homeowner may be trying to cover up some evident water damage.

Press down on the sides of the toilet bowl and attempt to rock it back and forth. If it moves with pressure, it may mean that the subflooring beneath the toilet has sustained water damage and has decayed. Ideally, the toilet should feel solid as the Rock of Gibraltar. Now lift the cover on the toilet tank and inspect the mechanism within. You'll be able to spot an old flush valve—replacing this someday will be a nuisance but no big deal. Newer low-volume toilets have pressure-assisted mechanisms. Because they use air pressure rather than sheer water volume to flush, they use much less water than standard toilets. This is generally a good sign.

Water pooling in a ceramic tile shower base means the tile is improperly sloped. This makes it likely that water will eventually penetrate through the grout and into the subfloor.

Laundry Room

In the laundry room, whether it's in the basement or a utility closet, make sure there is a floor drain or drain pan that will catch water if a leak should develop. At the laundry tub, look to see if the washing machine drain hose empties into a standpipe, or directly into the sink (a standpipe is preferable). Look to see if both the sink and standpipe drains are configured with standard drain traps.

It's a very common mistake for a laundry tub faucet to have an attached utility hose that lays directly in the bottom of the tub. If you find such a situation and end up buying the house, make sure you remove this hose or fit the faucet with a vacuum breaker (page 120). Unremedied, such a situation has the potential to allow dirty water to be siphoned back up into the fresh water supply, which could make your family sick.

It's also quite common for utility sinks in laundry areas to have "S"-shaped drain traps rather than "P" traps. S-traps are forbidden by plumbing codes, since they are more prone to siphoning, in which air pressure can suck the standing water out of the trap, thereby opening the home to rising sewer gases. Run water in the sink and listen for gurgling. This is a sign that the drain may be siphoning water from the trap. This is a situation you'll want to have corrected if you buy the house. Improving the venting of such a fixture can cost several hundred dollars.

For a laundry utility sink, proper plumbing includes a standpipe drain for the washing machine drain hose, as shown in this cutaway.

If a utility hose is attached to the laundry sink faucet, the spout should be fitted with a vacuum breaker that prevents water from being back-siphoned into the fresh water supply. The device costs just a few dollars, and can prevent some pretty severe health problems.

There's really no effective way for an amateur to assess a well; even accredited home inspectors usually waive on this one.

Sewer/Septic Tank

In rural areas, it's much more likely that the drain system will be serviced by a septic tank system rather than by a central sewer main.

A septic system consists of an underground tank and system of pipes fanning out from the tank. It works by draining or pumping solid and liquid waste out to the septic tank, where the solids settle to the bottom and liquids continue out to the perforated pipes in the drain field. You can identify a home serviced by a septic system first of all by the raised air vents out in the yard, or by the raised plateau of the drain field. You may also notice the presence of a waste pump inside the house, by which the waste is transported out to the septic tank. If the system is properly installed, it's pretty foolproof, requiring only that the pump be maintained, and the septic tank be pumped out by a service company every few years.

If the system has an extractor pump that pushes waste out to the septic tank, it typically lasts about 10 years, and you can expect to pay $400 or more if it needs to be replaced. This is an expected expense, but if you learn that the pump was already replaced in the last year or so, you will know that you should have a fairly long grace period before it's necessary again.

Ask the homeowner for records that show how often the septic tank was pumped. If this maintenance has been neglected, you could be looking at a full tank replacement, which can easily cost $5,000 or more.

Wells

If you're looking at a home in a rural area, there's a good chance that the fresh water supply will come from a well rather than from community water mains. You can identify a house served by a well by the presence of a large holding tank in the basement or utility room. There's really no effective way for an amateur to assess a well. Even accredited home inspectors usually waive on this one, recommending that a prospective buyer talk to a qualified well installation contractor. Ask to see any paperwork and maintenance records the homeowner has on the well. Well pumps typically have a life expectancy of about 10 years and cost between $500 and $1,200 to replace. If you look closely around the pump, you may see service tags than indicate the last time the well pump was serviced.

Plumbing the Depths: Plumbing System Checklist

Item	Good Indications	Good	Average	Poor
		Condition		
Water service	• ¾"or 1" entry pipe	☐	☐	☐
Water supply	• Copper, not plastic or iron	☐	☐	☐
Drain pipes	• PVC or copper, not iron	☐	☐	☐
	• Accessible main drain cleanout in utility area	☐	☐	☐
Vent system	• No sewer smell, or sounds of gurgling drains	☐	☐	☐
	• Visible vent stacks on roof	☐	☐	☐
Utility hose bibs	• At least two outdoor spigots, with shut-offs	☐	☐	☐
Water heater	• Proper water supply fittings	☐	☐	☐
	• T & P valve present	☐	☐	☐
	• Proper venting (gas)	☐	☐	☐
Kitchen	• No signs of leaking water	☐	☐	☐
	• Shutoff valves present	☐	☐	☐
	• Adequate disposal, dishwasher	☐	☐	☐
Bathroom	• Toilet secure, stable	☐	☐	☐
	• Well-sealed tub, shower seams	☐	☐	☐
	• Solid tile installations	☐	☐	☐
Laundry	• Standpipe drain	☐	☐	☐
Septic system	• Records showing service	☐	☐	☐

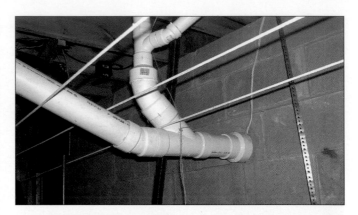

You can identify a septic system by the main drain that exits the house high on a basement wall (right, top), and by the presence of vent stacks above the drain field in the yard (right, bottom) Ask questions about the maintenance schedule when considering a home with a septic system.

If You're Selling:

You'll never recoup the investment of making major plumbing upgrades when you sell your house. It's much better just to adjust the sale price to create an allowance for the new owner to make these improvements. But it can be a great idea to invest a few hundred dollars in gleaming new faucets and other fixtures to help your house show at its very best.

Anatomy of a Home Plumbing System

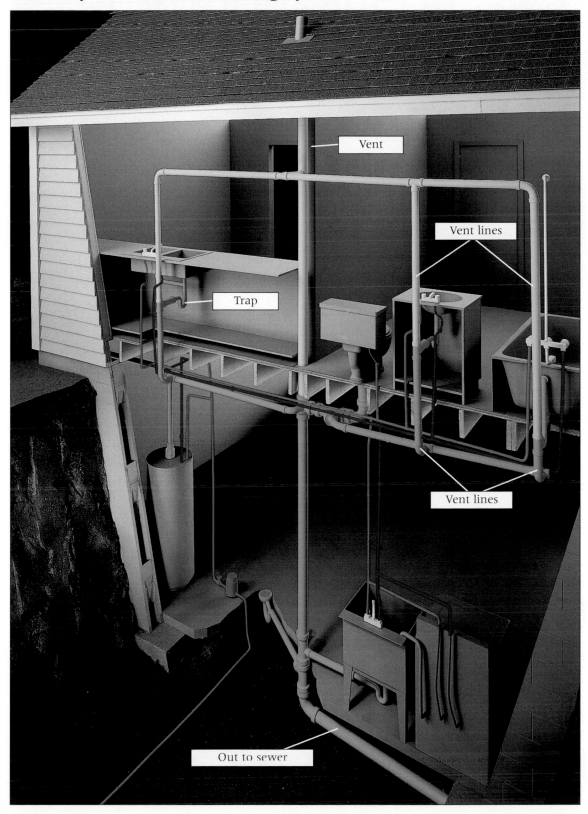

A plumbing system consists of two systems: the water supply system (red and blue pipes), and the drain-waste-vent system (drain pipes shown in green, vent pipes in yellow).

Vent

Vent lines

Trap

Vent lines

Out to sewer

Chapter 14

Of Ovens and Icemakers

Inspecting the Appliances

Photo courtesy of GE Consumer Products

Relatively new kitchen appliances can make a house worth $10,000 or so more than one with appliances that will soon need to be replaced.

Typically, if the appliance is built in, bolted down, or attached to a gas line, it is automatically part of the house.

Most prospective buyers gloss over the appliances when evaluating a home. After all, the appliances aren't nearly as glamorous as other features of the house, and in the greater scheme of things, they're usually viewed as being somewhat mundane. But it's not mundane if you have to replace all of them within a year—a typical house contains $10,000 to $20,000 worth of appliances. For this reason, make sure you give them some attention during your inspection.

Some real estate agents recommend that homeowners completely update their kitchen appliances before listing, and include them as part of the deal. This strategy is more likely to be used in expensive, upscale homes in a soft real estate market.

In some parts of the country—my part of the Midwest, for example—it's customary to include freestanding appliances as part of the package, while in others it is not. Certain appliances are considered "real" property, and are therefore included with the house. Typically, if the appliance is built in, bolted down, or attached to a gas line, it is automatically part of the house. It is, of course, always important to ask and to then list those appliances you want to include. Some built-in appliances may be included in a warranty, but freestanding appliances generally are not.

Inclusion of freestanding appliances may be a negotiating point, depending on whether it is a buyer's or seller's market when you're buying. Obviously, if houses are being sold within hours of listing, the appliances are not an issue. However, in a buyer's market, a seller may respond to a low offer by including appliances as part of their counteroffer.

If the appliances are included, what you want to know is their average life expectancy so you can budget for replacement. The chart on page 125 shows the average useful life for common appliances.

How do you determine the age of appliances that are coming with the house? Some owners may have sales receipts that show when the appliances were purchased. If not, look for the manufacturer's tag that indicates the age of the appliance, and use the chart to estimate how much useful life is left in each appliance. I prefer that appliances have at least half their useful life remaining, whenever possible.

Take into consideration the fact that older appliances can be energy hogs, and consequently may be more costly to keep than to replace. That 25-year-old harvest gold fridge may still be chugging along—but you'll be paying at least an extra $100 a year in energy costs for it. In some parts of the country you will want to look at the water use of dishwashers and washing machines, as well. Older models will not have as many conservation features.

The condition of appliances can give you insight into the general level of care the homeowner has invested in the home. Is the dryer lint trap full to bursting? Is the oven caked with burned-on drippings? Many sellers will conscientiously clean the obvious parts of their home but not realize the nitty-gritty secrets revealed by a grimy refrigerator bin.

When you decide to buy, test the included appliances. Bring along thermometers for checking the fridge, freezer, and oven. Make sure they operate at the tempertures to which they're set. Turn on all individual burners or cooking surfaces. If the stove or dryer are gas powered, smell for gas odor when you turn them on. Run the dishwasher, garbage disposer, exhaust hood, vent fans, and trash compactors.

APPLIANCE	AVERAGE USEFUL LIFE (years)
Room size air conditioner	10
Dishwasher	10
Dryer	14
Electric range	17
Garbage disposer	10
Gas range	19
Microwave	11
Refrigerator	14-17
Washer	13
Trash compactor	14
Dehumidifier	11
Front load washer	11
Freezer (upright or chest)	15-18

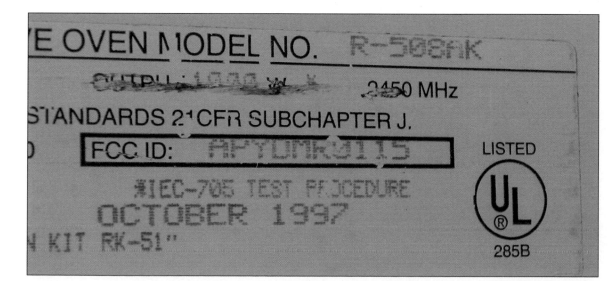

A manufacturer's tag should be present on every appliance, though you may have to search for it. The tag indicates when the appliance was manufactured. Remember, though, that the actual purchase date was probably a year or two later than the manufacturer date.

Signs of Gold:

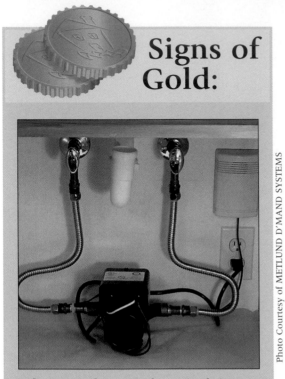

A hot-water heater with a recirculating pump delivers hot water to faucets when they are turned on, so you don't have to run the water until it becomes hot. When you turn on the water there is a one- or two-second delay as the cool water in the pipe is returned to the hot-water heater and hot water is delivered to the faucet. This is important in areas that have limited water supplies, as 1 to 2 gallons of water can be wasted each time the user waits for hot water.

Photo Courtesy of METLUND D'MAND SYSTEMS

Refrigerators make noise as they run—you can often hear the refrigerant percolating, the fans running, and the drip of water as the freezer automatically defrosts. If you can hear these sounds from another room, however, the cooling unit may be at the end of its useful life.

Many appliances, such as refrigerators, washing machines, dryers, and window air conditioners, have bearings or belts, which should not squeal. Squealing is a sign that the motor bearings are dry or nearly dry and that the motor will burn out sooner rather than later.

A garbage disposal should have at least a one-half horsepower motor. If not, it will be subject to clogs or motor burnout. Check the horsepower, which is marked on the housing of the disposal.

Make sure that you don't get stuck with any appliances you don't want. That old chest freezer in the basement will be expensive to have hauled away if you don't want it. Any unwanted items should be mentioned in the purchase agreement so the seller is responsible for disposal.

A faulty door seal has caused this refrigerator to rust from condensed moisture.

If You're Selling:

Generally, the more appliances you include in the sale price of the home, the better—provided they are in good operating condition. One way to reassure prospective buyers is to take out insurance on your appliances and pay for the first year or two of premiums. These policies cost only a few dollars, and will reassure your buyers that they won't be faced with exorbitant replacement costs in the immediate future. Contact companies that specialize in home warranty insurance.

Clean all appliances very thoroughly. The more pristine the appliances appear, the better your house will show. A grimy oven will turn off a great many buyers.

Of Ovens and Icemakers: Appliance Checklist

Item	Good Indications	Condition		
		Good	Average	Poor
Stove/Range/ Cooktop	• Less than 8 years old • Plenty of burners; adequate oven space	☐ ☐	☐ ☐	☐ ☐
Refrigerator	• Less than 7 years old • Energy-efficient model • Includes water dispenser, automatic icemaker	☐ ☐ ☐	☐ ☐ ☐	☐ ☐ ☐
Microwave oven	• Less than 5 years old • Ample size	☐ ☐	☐ ☐	☐ ☐
Dishwasher	• Less than 5 years old • Energy-efficient model	☐ ☐	☐ ☐	☐ ☐
Freezer	• Less than 8 years old	☐	☐	☐
Trash compactor	• Less than 7 years old	☐	☐	☐
Food disposal	• Less than 5 years old • ½ horsepower, minimum	☐ ☐	☐ ☐	☐ ☐
Washing machine	• Less than 5 years old	☐	☐	☐
Clothes dryer	• Less than 7 years old	☐	☐	☐

Chapter 15 On the Surface of Things
Walls, Ceilings, Floors, & Other Surfaces

Surfaces can make a huge difference in the appearance of a home, as shown in this kitchen makeover. Don't overlook a home with dated surfaces (below), which can be quite easily transformed with new surfaces (above).

When it comes to the surfaces in a home, there's bad news and there's good news. First the bad news. A home with dreadful carpet, wallpaper, or paint can look just plain awful when you inspect it. It's hard to see past burnt orange shag carpeting dating from the 1970s. (I know, because I bought a home that had this exact floor covering.) You do need to consider the cost of these changes as you evaluate the home; even cosmetic changes cost money. Too many "minor" cosmetic issues can add up to an expensive headache. Also, consider the fact that despite whatever plans you have for redecorating, you will no doubt have to live with the previous owner's taste for some time before you get things changed.

But resist the urge to run screaming into the night just because the surfaces aren't exactly your cup of tea. The good news is that these things are generally no big deal. If you have the discipline necessary to overlook cosmetic issues, you might get yourself a bargain when it comes to buying a home. Lots of people will be turned off by a superficial unattractiveness that disguises a good deal.

Many professional real estate investors know this truth: redeeming a house with bad decorating is relatively easy, provided the house is sound when it comes to the important features described elsewhere in this book. A house with cosmetic inadequacies may sell for tens of thousands of dollars less than a comparable house decorated with a few hundred dollars in new wallpaper and paint.

Best of all, redecorating is a project that can be tackled room by room—unlike a bad roof, for example, which may require that you shell out $15,000 or $20,000 instantly.

Water stains on a ceiling around a fireplace chimney are likely caused by a roof or flashing problem.

This ceiling stain is beneath a bathroom plumbing fixture—reason to investigate further.

While I encourage you to look past carpeting, wallpaper, or paint that is simply ugly, I do want you to look closely at all these surfaces to make sure there aren't serious problems. In particular, problems with walls, ceilings, and floors can hint at more dramatic flaws in plumbing, roofing, or structural systems.

Walls & Ceilings

As you enter a room, scan the condition of the walls. Small, hairline cracks in corners are a normal sign of house settling. Large cracks (greater than 1/16 inch wide), although rare, can be a red flag for possible structural issues with the house. Also beware of any cracks that run the length or width of a wall or those that extend from walls up onto the ceiling. A professional inspection should be performed if you note any cracks like these.

Drywall. Inspect walls for popping nails or badly taped joints. These are not major issues but can be signs of shoddy workmanship, so be on the lookout for other signs around the house. Rust stains around drywall fasteners or corner beads indicate moisture damage. Pay especially close attention in the basement and beneath any windows, as these are the usual locations of water intrusion in a home. Because the finished walls in a basement are a few inches away from the foundation walls, signs of water damage generally won't appear on the middle and upper portions of the walls. Instead look along the walls near the floor for signs of water damage.

A water-damaged ceiling in this closet was eventually traced to an ice dam on the roof.

Discolored woodwork, especially along baseboards, can indicate a moisture problem you shouldn't overlook.

Peeling wallpaper is often a sign of a house with ventilation problems. Humid air that's not properly vented causes this problem.

Plaster. Plaster walls may be cracked or disintegrating, or may have missing sections. Gently tap and push on plaster; if any area sounds hollow or feels flexible, it's a sign that the plaster has separated from its backing. Bulging plaster can be a sign of structural issues.

Trim. Inspect the trim along both the floor and ceiling for any cracked, broken, or missing pieces. These are usually just cosmetic issues, but if you notice any gaps between the walls and ceiling or separation between walls of greater than ¼ inch take note, as these can be warning signs of larger structural problems.

Paint. On painted walls, look for peeling paint or discoloration from moisture damage. Old, peeling paint can also be a main source of lead poisoning. Older paint jobs are quite likely to contain lead, but unless tested, it is impossible to know for sure. This condition is potentially very dangerous and should be dealt with as soon as possible. Also be aware that if you are applying for an FHA loan, the federal government may require that any peeling paint be removed or sealed over before the closing.

Wall covering. If the walls are wallpapered, examine the seams, since that's where wallpaper usually begins to peel. This isn't a big deal by itself, but if you notice large, loose sections, probe carefully behind the wallpaper for any signs of mold or moisture damage. Wallpaper sometimes disguises wallboard or plaster that is badly disintegrating. Taking down wallpaper in such situations can cause the walls to literally come apart. Sometimes vinyl wallpaper acts as a moisture barrier over drywall, creating damp conditions that foster mold growth.

Paneling. Sometimes paneling is attached without any sort of backer, like drywall, behind it. Without backing, thin paneling may warp between the studs. Tap along the paneling and listen for a solid sound to see if backing is present.

Textured ceilings. Notice whether the rooms have "popcorn" ceilings. These spray-on ceilings have a heavy texture that is more pronounced than newer textured ceilings. They were very popular in the 60s and 70s. If you like the look and there are no damaged areas, there's no problem. Pay attention if you see damaged or flaking areas, though. Tests may reveal that the popcorn spray contains asbestos (see Chapter 17). If asbestos is present, plan to pay at least $5 per square foot to have it removed.

Tiled ceilings. There are several different types of tiles used on ceilings, including suspended acoustic tiles, adhesive-mounted fiber tiles, and even metal tiles. It's easy enough to judge the condition of a tiled ceiling—turn on the ceiling fixtures and look carefully for signs of buckling or bowing, which would tell you there might be underlying structural problems. With suspended ceilings, lift up a few panels and look up in the ceiling cavities to judge the condition of the joists and cross-blocking.

Flooring

Carpeting. Your main concern with carpeting should be its condition and quality. All carpet will eventually show wear in high-traffic areas, but these areas may not be noticeable if the owner has rearranged the furniture to adjust the traffic in a room. While it's generally not acceptable to move a seller's furniture around during a viewing, you can ask the owner about any unseen damage to the carpets.

A good way to evaluate carpet is to remove your shoes and see how it feels beneath your feet. A quality carpet will have good padding underneath. Carpet with a solid, thick pad beneath will feel thick and lush. If the carpeting feels hard or mushy underfoot, the padding is too thin or too soft. You can also compare a carpet's quality by getting down on the floor and taking a close look at the strands of carpet fiber. Higher-quality carpets usually have a thicker pile. Cheap carpeting spreads easily, revealing the backing beneath.

Average Installation costs

Walls & Ceilings	
Install new drywall	$3.50/sq. ft.
Repair plaster	$75/sq. yd.
Paint walls	$200/room
Paint ceilings	$125/room
Install acoustic ceiling	$5/sq. ft.
Install ceiling tile	$2.75/sq. ft.
Remove wallpaper	$.75/sq. ft.
Hang wallpaper	$1.25/sq. ft.

Floors	
Carpeting with pad	$25 to $75/yd.
Hardwood plank	$15/sq. ft.
Sand & refinish	$1.75/sq. ft.
Parquet wood	$11/sq. ft.
Sheet vinyl	$5 to $7/sq. ft.
Ceramic tile	$14/sq. ft.

Cabinets & Countertops	
Replace base & wall cabinets	$250/lin. ft.
Refinish cabinets	$85/lin. ft.
Laminate countertop	$40/lin. ft.
Ceramic tile countertop	$85/lin. ft.

A good way to tell a quality carpet is to remove your shoes and see how it feels beneath your feet.

Signs of Gold:

A well-installed ceramic tile floor is an elegant, durable surface that is a notch above any other type of flooring in a kitchen or bathroom.

Hardwood floors may lie under the carpeting, giving you a great remodeling option. To determine the type of flooring that lies beneath the carpet in the house you're inspecting, check closets. If you see a hardwood floor in the closet, it probably means the entire house has hardwood floors. Another way to check flooring is to remove floor grates or cold air returns, where you'll be able to see the edge of whatever flooring lies beneath the carpeting.

Wood floors. If you walk into a home and see rooms full of wood floors, make sure you know exactly what you're getting. Synthetic wood floorings (such as Pergo) are very popular and sometimes difficult to distinguish from true hardwood. Tap on the floor. Hardwood makes a dull thick sound. Synthetic wood flooring, because it actually floats on the underlayment, makes a hollow, reverberating sound.

It's also possible you may come across floors made from softwoods like pine, fir or cypress. Often these floors have a rough surface and provide a rustic, "cabinlike" feel to rooms. Be aware that softwood floors need to be resanded every 5 to 10 years. Make sure there is enough wood thickness left to resand.

Note the general condition of the wood floors. Are they scratched? Do you notice any warping? Look for any popped nails. Walk around on the floors and you'll easily feel the movement of warped floor boards. Hardwood floors in potentially wet areas like kitchens, bathrooms, or laundry rooms are often the site of warping. Severely damaged floors can cost upwards of a few thousand dollars to entirely replace, depending upon the area.

Ceramic tile floors. A ceramic tile floor is generally a very good floor, provided it was installed correctly. However, if you see a ceramic tile floor with many cracked joints and obvious buckling, it's a sign that it was probably installed over a thin plywood subfloor, rather than a layer of cementboard or poured mortar. A bad ceramic tile floor in a large kitchen can cost thousands to remove and replace.

If the grout joints are badly discolored, you may need to have the floor regrouted—a job that will cost several hundred dollars per room.

Vinyl flooring. As with all flooring materials, note the general condition of vinyl flooring, looking for major tears or damage. Small cuts and scratches are easily reparable, but if the flooring is badly worn or if the damage is widespread, the only option is replacement.

Cabinets & Countertops

Cabinets and countertops are the last major surfaces in a house, and have a major impact on the look of the home. Aesthetically, you'll know instantly if you like what you see, but here are some things to look for:

- Good cabinets have solid hardwood drawers and doors rather than plywood or particleboard.
- On laminate countertops, look for signs of water damage around the sink at seams between the segments of countertop.
- Ceramic tile, granite, or solid-surface materials (such as Corian) are premium signs of gold when it comes to countertops.
- Make sure both countertops and cabinets are securely fastened.

Tracks of the Poor Craftsman:

Skilled installers always remove stop moldings and baseboard shoe moldings before laying ceramic floor tile. The moldings are then reinstalled after the floor is laid. When you see a ceramic tile floor where the tiles have been cut to fit around moldings, it means that the installer was inexperienced. Always inspect more carefully when you see signs of an inexperienced craftsman.

If You're Selling:

- Remove wallpaper, especially if it has a dramatic, loud pattern. Bold wallpaper is often a turn-off to prospective buyers.
- If you're repainting, use neutral tones, which make spaces seem larger and rarely offend anyone's sensibilities.
- A good investment is installing custom wood moldings, such as a chair rail molding or crown molding (right). A good-quality molding is an inexpensive way to make your home look a lot better.

On the Surface of Things: Walls, Ceilings, & Floors Checklist

Item	Good Indications	Condition		
		Good	Average	Poor
Walls	• Solid, no signs of water damage	☐	☐	☐
	• Smooth, no severe cracking	☐	☐	☐
	• Wallpaper pleasing	☐	☐	☐
Floors	• Carpet of good quality	☐	☐	☐
	• Vinyl flooring solidly bonded	☐	☐	☐
	• Wood floors have no scratches, cracks	☐	☐	☐
Cabinets	• Solid hardwood doors	☐	☐	☐
	• Aesthetically pleasing	☐	☐	☐
Countertops	• Aesthetically pleasing	☐	☐	☐
	• Secure, no damage	☐	☐	☐

Anatomy of Walls, Ceilings, & Floors

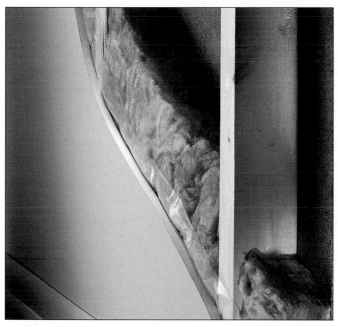

Wallboard panels are generally screwed or nailed directly to studs. Look for traces of rust over screw heads or nail heads, which tell you the wall allows moisture to penetrate.

Plaster is applied in layers over wood, metal, or "rock" lath. Look for areas of buckling, and tap on surfaces to listen for areas that sound hollow.

A typical floor is supported by both a subfloor and an underlayment. Without both layers, a floor is prone to flexing, sagging, and unevenness.

Suspended ceilings are often installed in basements to provide space for recessed light fixtures.

Chapter 16

Nice to Meet You
Meeting the Owner

Time of year can have a dramatic impact on the look of a house. Family photos reveal that this house has a beautiful landscape (above) that would be almost impossible to imagine if you happen to be inspecting in late winter (below).

Surprisingly often, prospective buyers never get to talk to the owners of a home at all before making an offer on the property. My suggestion here is a somewhat controversial one, but I strongly recommend you spend at least an hour or so with the owners at some point after you've done your visual inspection. This meeting can provide key information that convinces you to either put down an offer or look elsewhere. I know of several instances where owner and buyer became longtime friends after such a meeting.

It's possible that you'll meet the owner as part of your first inspection visit, but it's more likely that you'll meet during a follow-up visit. How you arrange this meeting will depend on how the home is being sold.

FSBO Homes

If the open house is a *For Sale by Owner* (FSBO) affair, it's very easy to meet the owners, because they're the folks showing the home. If you're viewing the home at a public open-house showing and you find yourself seriously interested in the property, it's a good idea to stick around until the closing hour of the event. As other visitors leave, ask to sit down with the owner for a few minutes to talk in depth about the house. If you're seriously interested, the owner will be quite happy to talk to you at length. If there are many people still milling about, though, it might be best to arrange a separate meeting a short while later.

Agent-represented Homes

When the house is listed by a real estate agent, it can be a bit more difficult to meet the owners. This is where my advice gets a bit controversial. Real estate agents generally discourage owners and prospective buyers from meeting before the purchase agreement is completed. In some cases, there is an almost religious fervor about forbidding this kind of meeting, especially one that takes place without the agent's knowledge. From the agent's perspective, this is mostly about protecting their financial interests—they don't want to take any chance of losing the sale or their commission. Since you have no such ulterior motives, don't be daunted by this taboo. Make sure everybody knows you're not attempting to bypass the agent, and find a way to meet the owner. If there is no other way, talk to the agent about arranging a joint meeting including all three parties: you, the owner, and the agent.

I've sometimes found it possible to talk to the owner by stopping by the house an hour or so after a public open-house showing ends—after the agent is gone, but while the owners are still in open-house mode. If that's not practical, you might ask to stop by for a few minutes on a Saturday afternoon to look at the trees and flowers. In many cases, a friendly conversation under such circumstances will work into an invitation to have a cup of coffee and a more in-depth discussion.

After You Say Hello

Be sensitive to the owner. Be aware that any meeting with a prospective buyer feels stressful to owners, so don't take the nervousness personally. And do whatever you can to put them at ease. Meet the owners on their own terms—you can simply offer to treat them at a local coffee shop if they're hesitant to meet privately at the home. In all likelihood, the owners have spent many hours preparing the home for sale, and any additional demands on their time will be burdensome—so go out of your way to make it easy. Finally, if the owners are very reluctant to meet with you, you'll just have to respect their wishes. If you have important questions, have the real estate agent run interference and get the answers you need.

When you do meet the owner, structure your conversation as though you're planning to move into the house. If you've reached this point, the house is on the short list, and the owner will be put at ease if he or she understands that you're now a very serious shopper. Start by talking up the things you like about the house.

It's always good to be genuinely interested in the owner's lifestyle and family. Almost everybody enjoys talking about their family, and before you know it, a whole history of the home is unfolded for you. You might learn, for example, that the basement rec room was once a drive-under garage that was converted when the oldest son reached his sixteenth birthday, which is also the date

Room deodorizers tell you that an unpleasant odor could be hidden. Try to find out what this is about. Causes could include mold due to water damage or pet accidents.

Almost everybody enjoys talking about their family, and before you know it, a whole history of the home is unfolded for you.

Signs of Gold:

A neatly organized workshop with all the typical maintenance products—caulks, sealers, paintbrushes, and the like—suggests that the owner has been diligent in the upkeep of the home. This owner might be a tad compulsive, but that's not a bad thing when it comes to home maintenance.

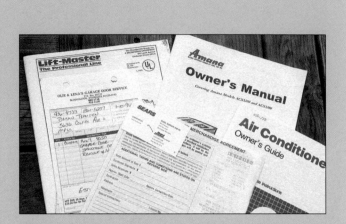

A homeowner who has has kept all receipts and warranty information (above) for the appliances has probably kept them in good working order. When we visited the owner of our current home, they showed a complete file of this kind of information.

that the double-car garage was added out back. Now you know important dates for major remodeling projects in the home.

At one sit-down meeting, an owner pulled out family photos that gave me glimpses of the home at various key points in its history—including the storm that dropped a huge tree limb and marked the date the home was reshingled.

Share and share alike. Share some details of your own life—this puts owners at ease and makes it likely they'll share things with you. If you feel comfortable doing so, ask why the owner is moving. A home is sometimes underpriced if an owner needs to sell quickly—because of a sudden employment change, for example, or a divorce. Often in these instances, the owners don't have time to invest in cosmetic improvements, so the house may not "show" quite as well as it might. This means you might face less competition for the house.

Once everybody is comfortable, ask any pertinent questions about the age and condition of appliances, the roof, the siding, and anything else that draws your attention. If you didn't get to see the garage, attic, or utility spaces during the official open house, ask to see these areas now. You might run into some hesitation about this—but be firm. You really do need to see these areas before you commit to a purchase.

If, during your primary inspection, you found areas of concern, ask about them now. And ask about the completion dates for any major remodeling projects. If the sale price includes the appliances, ask about their ages and warranties. When I asked one homeowner about this, she produced a large file full of information on cleaning and service dates for all the home's utilities—sewer, furnace, air-conditioner, and all the rest. It was very valuable information that convinced me the home had been kept in good working order over the years.

Questions for the Owners:

- How long have you owned the house? Do you know anything about the previous owners?

- What are the good features of the neighborhood? Schools? Parks? Restaurants?

- Where do favorite neighbors live? Least favorites? Any tips for getting along with them?

- How has the neighborhood changed? Are the residents predominantly families with kids? Empty nesters?

- What are the detail of civic services—snow plowing, trash and recycling collection, etc.?

- What species of trees, shrubs, and flowers does the yard have? When were the trees last trimmed?

- Was the lawn maintained by a lawn service? If so, would the owner recommend it?

- When was the house last roofed? Is the paperwork available?

- When was the house last sided? Paperwork?

- When was the last sewer work done? Does the owner have a plumber who has provided regular service?

- Have any significant electrical work or upgrades been done on the home?

- Has the house had major remodeling that called for building permits? If so, when was this work done?

- Any major redecorating? New carpets? Wallpaper? Plumbing fixtures?

- How old are the appliances? Are there receipts and owner's manuals available?

- Have the owners maintained pets in the home? (Mentioning your own pets can be a good way to elicit this information.)

If You're Selling:

Don't be nervous if prospective buyers want to meet you. It means they are very seriously interested in your home. When you meet, be open and honest. It can be very reassuring to a buyer if you point out any problem areas and don't try to soft-pedal problems that are obvious. Your honesty will be recognized and appreciated.

Unless you're working independently as your own negotiating agent, don't enter into any price discussions at this meeting with a prospective buyer; defer all these questions to your agent.

Chapter 17

The Healthy Home

Avoiding the "Sick House"

Along with health dangers, home buyers need to pay attention to environmental issues because dealing with these problems is invariably expensive.

So you've found a home that fits you. No major structural problems, no water in the basement. The roof is in great condition; the electrical system is adequate; and the bright spacious kitchen is a dream. All that's left is to sign the papers and move on in...right? Not so fast, my friend.

In recent years concern over the health of a home's environment has become as important as a home's physical condition for many buyers, and rightly so. The term "sick house syndrome" refers to any of a wide variety of environmental issues that actually cause illness for a home's occupants. Issues of indoor air quality, water quality, and other environmental hazards around the home can have health consequences ranging from merely irritating to fatal. Along with the health dangers, home buyers should be concerned with environmental issues because dealing with these problems is unavoidable and often expensive. If problems exist, the cost of resolving them needs to be considered in the overall sale price of a home.

Unfortunately, home health problems can't reliably be diagnosed with a flashlight or a level over the course of a home inspection visit. Many require lab testing to identify hazards definitively. The fast-paced, competitive nature of the some housing markets may make it impractical to wait weeks or even days for test results before making an offer.

However, there are clues that can give you a heads-up to potentially dangerous and costly health concerns. When you're ready to make an offer on a house, refer to the the notes you've taken and have tests conducted as needed. Provisions for pending tests can be written into purchase agreements. Discuss these possibilities with your realtor and lawyer.

Health hazards aren't found only in substandard housing. Radon gas, carbon monoxide, poor water quality and other enemies can hide in the most luxurious of settings.

Indoor Air Quality

When you think about air pollution, images of belching factory smokestacks, choking traffic exhaust, or reeking landfills may come to mind. But most people don't think about the air inside the four walls of their home. In fact, studies have shown that the air inside some homes can be more polluted than what is permitted by federal regulations for outdoor air. Because people spend most of their time indoors, the quality of the air inside a home is a serious matter. Poor indoor air quality can have health effects ranging from mild headaches and eye irritation to serious organ damage and even death.

Ventilation is the key to indoor air quality. Insufficient ventilation allows the residues of cleaning solvents, cooking odors, smoke, and exhaled air to remain. Over time these polluting substances can reach irritating levels. Poor ventilation also allows moisture to build up in a house, which can lead to mold and bacteria problems, as well as decay in a home's physical structure. The energy efficiency of many of today's modern homes has the unintended side effect of sealing occupants in with possible contaminants. While I don't mean to suggest that you are better off buying a drafty house, the need for proper ventilation in a home can't be overstated. The best solution for all houses is an air-to-air heat exchanger, also called a heat recovery ventilator. This device pulls in fresh air and exhausts stale air without exhausting the heat or air-conditioning.

An air-to-air exchanger is an important component in maintaining good indoor air quality.

> Air-to-air heat exchanger: $2,000 to $2,500

Mold

Recently, the health effects of inhaled mold spores on humans have begun to gain public attention. Stories of schools, offices, and homes evacuated because of high concentrations of toxic mold spores have made headlines. Serious health problems ranging from memory loss and fatigue to brain damage and seizures have been blamed on so-called "toxic mold syndrome."

Although the scientific jury is still out on the seriousness of this problem, documented studies have shown that children in schools with high mold levels had more frequent asthma attacks and higher incidences of wheezing and upper respiratory infections. Since children spend more time at home than at school, it is safe to assume that high mold levels at home would have negative effects. For the 10 percent of the population with severe mold allergies, even a visit to a mold-infested house can be dangerous. Mold can

Mold not only looks (and usually smells) horrible, but it may well be consuming the wall and the framing behind it.

Home humidity should be at 50% or lower to prevent mold growth. Sweaty windows may be caused by high humidity or insufficient weather stripping.

also have serious effects for those with asthma or immunodeficiencies. But that doesn't mean that there is a reason to panic over the presence of mold.

Some mold will be present in virtually any house. Mold is a natural organism that consumes organic matter and speeds its decay. Mold becomes a problem when mold spores inside a home are supplied with sufficient moisture to continue consuming organic matter and begin to reproduce in an unchecked manner. Molds will eat away at the organic material in drywall, insulation, and wallpaper, or the glue that bonds carpets together.

Moisture is the key to encouraging mold, and the welcome mat is a damp, poorly ventilated house. Mold spores need moisture to establish themselves and thrive. That's why whenever you see evidence of moisture or water damage, you should be suspicious of mold growth as well. Look for water damage from leaking plumbing or a leaky roof. Ask about any previous water damage from flooding, leaks, or backed-up sewers. Often overlooked sources of moisture include crawl spaces without moisture barriers and condensation on uninsulated air-conditioning pipes or ducts running through attics. Finished basements harbor great mold potential due to the naturally higher humidity levels.

An obvious indicator of a mold problem is a musty odor. Visible signs include white thread-like growths or clusters of small black specks along walls. However, hidden mold can often be found growing behind wall coverings and ceiling tiles, or inside ductwork.

If you are looking at a newer home with a stucco exterior, be aware that there have been severe mold problems in houses that have an improperly applied Exterior Insulation and Finish Systems (EIFS) exterior. Referred to as synthetic stucco, this very popular exterior has been used extensively over the last 10 years or so. Problems occur when the joint areas between the stucco and roofing or windows are not properly sealed and moisture seeps behind the exterior. A faulty installation costs thousands to repair—plus the likelihood of a severe mold infestation is quite high due to the continual moisture influx.

Why does mold need to be removed once the moisture source is eliminated? The mold spores are the toxic elements. A mold colony releases spores when physically disturbed, even if the colony is dead. A severe mold infestation also destroys the structural integrity of carpet, drywall, and insulation.

People are paying tens of thousands of dollars for professional removal of large mold infestation from their homes. Major lawsuits and insurance claims for mold infestation have raised insurance premiums in many parts of the country and many homeowners are no longer able to purchase mold or water damage coverage. So be careful about buying a house that poses serious mold problems.

Mold remediation: $10 to $150 per square foot

Radon

Radon is a naturally occurring radioactive gas formed from the decay of uranium or radium in the soil and rock under a house. Until recent years, radon gas was probably not the sort of thing people gave much thought to, but in the mid-1980s scientists began to recognize and call attention to the potential dangers of radon in the home. As radon gas moves up through the ground, it can enter your home through cracks and other holes in the foundation, where it can become trapped and build to dangerous levels. Since radon is invisible, tasteless, and odorless, you may never know you are being poisoned.

The Environmental Protection Agency (EPA) estimates that 1 out of every 15 homes in the United States may have elevated radon levels. According to a National Academy of Sciences report, radon is the cause of between 15,000 and 22,000 lung cancer deaths each year. In fact, radon is second only to smoking as a leading cause of lung cancer. Combine high radon levels in your home with someone who smokes and the cancer risks multiply. Of all the potential health hazards present in homes, radon poses the greatest risk of serious long term illness.

The EPA has developed maps that predict the potential indoor radon risks for every county in the U.S. While these can give you a general sense of an area's risk potential, they won't tell you the radon risk of any

A basement crawl space with a dirt floor and no vapor barrier can be a source of moisture and mold, and also radon gas.

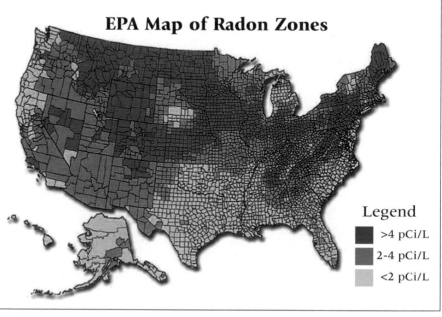

EPA Map of Radon Zones

Legend

■	>4 pCi/L
■	2-4 pCi/L
■	<2 pCi/L

The EPA has developed maps based on geological data that predict the potential indoor radon risks for every county in the U.S.

Radon levels in a home depend on a wide range of factors, including the construction of the house, the uranium content of the soil, and the geologic formations beneath the house. For this reason, you can't rely on test results for a neighbor's home.

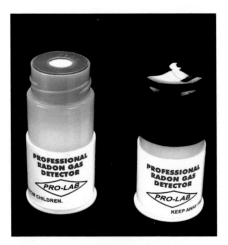

Radon test kits are available, but must be sent to a laboratory for analysis.

particular house. Any home can have high radon levels, even those in low radon-potential zones. Radon levels in a home depend on a wide range of factors, including the construction of the house, the uranium content of the soil, and the geologic formations beneath the house. And you can't rely on the results for a neighbor's home for guidance; even homes next door to each other may have dramatically different radon levels. The only way to know a home's radon level is to test.

Long-term tests that compile the average radon level for a home are the most accurate. However, you can demand that a short-term radon test be performed to give you a snapshot of the home's radon levels. Satisfactory results of such a test can be a condition of sale. A state or local radon official can recommend a test device most appropriate to your needs and testing conditions. These devices are generally placed on the lowest level of a home suitable for occupancy, exposed to the air for several days, and then sent to a laboratory for analysis. If you or the homeowner perform the test, be sure to follow the testing procedures precisely to ensure accurate results.

The concentration of radon in the air is measured in units of picocuries per liter of air (pCi/l). Based on the most current information, the EPA has set the "action level" for radon in homes at 4 pCi/l; that is, no action is need if the radon concentration is below 4 pCi/l. The EPA admits, however, that even this level is not risk free; annual exposure to 4 pCi/l is equivalent to the risk of smoking 10 cigarettes a day or having 200 chest X-rays a year.

If your test shows elevated radon levels, there are several

EPA Guidelines for Radon

Annual Average Radon Level	Recommended Action
Over 200 pCi/l	Take action within several weeks to reduce levels
Between 20 and 200 pCi/l	Take action within several months to reduce levels
Between 4 and 20 pCi/l	Take action within several years to reduce levels to about 4 pCi/l (sooner if levels are at the upper end of this range)
Below 4 pCi/l	Although there is still some cancer risk, reduction below this level may be difficult or impossible to achieve
Below 1.0 pCi/l	Average first-floor residential levels

methods for reducing the levels. Ventilation alone does not solve the radon problem. Ideally, the radon needs to be vented from the ground underneath the house. A pipe and fan system that vents radon gas from below the floor slab to above the roof is the best radon-reducing procedure. Make sure you have an EPA-approved radon-mitigation contractor do the work; correcting a radon problem is not a job for a do-it-yourselfer. Be sure to keep documentation of radon reduction improvements as a selling point for the home in the future.

Radon remediation: $500 to $2,500

Asbestos

Asbestos is a naturally occurring, fibrous mineral found all over the world. Its fibers are strong, fireproof, corrosion-resistant, and excellent insulators. It was these properties that fueled its use in construction products such as roof shingles, textured paints, insulation, and flooring tile. Unfortunately, despite its usefulness and versatility, asbestos causes fatal lung disease in those who live and work around it.

Tiny asbestos fibers are easily inhaled into the lungs, where they scar the lung tissue and cause several types of cancer. According to the EPA, between 3,000 and 12,000 people die each year from cancer related to asbestos exposure. Due to the health concerns and potential liability, most of the asbestos materials that had been used in construction are no longer manufactured. However, various asbestos-laden products can still be found in many homes built before 1978.

Asbestos fibers can be positively identified only with a microscope, but you can identify some of the more common materials that probably contain asbestos. For example, the heating pipes in many older homes are often covered with insulation that contains

The Environmental Protection Agency estimates that 1 out of every 15 homes in the United States may have elevated radon levels.

Common Asbestos Problem Areas

Attic and Ceiling

- Insulation
- Overhead pipes
- Flues and chimneys
- Sprayed ceilings

Living Quarters

- Floor tile
- Overhead pipes
- Flues and chimneys
- Sprayed ceilings

Basement/Utility Area

- Floor tile
- Overhead pipes
- Flues and chimneys
- Sprayed ceilings

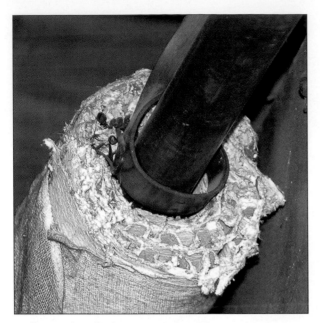

Asbestos insulation was commonly applied to heating pipes in older homes. It has a corrugated cardboard-like appearance.

Removal of asbestos is not always necessary and, in fact, is often the least desirable alternative.

asbestos. As you walk around the house, look for possible asbestos problem areas. Inspect potential asbestos sources for loose, torn, or damaged sections. Make note of the potentially hazardous areas. You might also ask the homeowner if he or she has already had any of the trouble spots tested for asbestos. Ask to see copies of the test results.

Despite its dangers, the mere presence of asbestos in your home is not necessarily hazardous. The danger lies in asbestos materials that may become damaged over time and release asbestos fibers into the air. Friable material (material which can be crumbled by hand) is of particular concern. However, the risks of exposure are small if asbestos material is in good condition. The best solution is simply to leave it alone. Disturbing asbestos can create a hazard where there was none before. Removal is typically very expensive and has the most potential for releasing fibers into the air. Depending upon its condition, a professional may recommend encapsulating the asbestos material with a special sealant to prevent the fibers from being released or enclosing the material completely to contain the particles.

Another possible asbestos source you may be able to recognize is vermiculite insulation. Vermiculite is another naturally occurring mineral that often contains asbestos and was sold as attic insulation until the mid-1980s. It is easy to distinguish vermiculite from other more common types of loose-fill insulations, such as cellulose, fiberglass, or rock wool, as these are all fibrous. Vermiculite comes in the form of brownish-pink or brownish-silver accordion-shaped chips. As long as the insulation is left alone, the health risks from vermiculite are small. Simply being aware of its presence will allow you to take proper precautions when working around vermiculite.

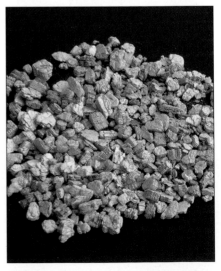

Vermiculite comes in the form of brownish-pink or brownish-silver accordion-shaped chips.

Asbestos removal:
$1,500 to $3,000
Asbestos encapsulation:
$2 to $6 per sq. ft.

Lead

Throughout the nineteenth and early twentieth centuries, house paints were laced with high amounts of lead, often as much as 10 percent and sometimes as high as 50 percent. Lead made paint more durable, so a high lead content signified a high-quality paint. In addition to paint, lead was used in hundreds of products, from water pipes and gasoline to ceramic dishes and food cans. And then, as with many things that "seemed like a good idea at the time," came the bad news: lead is extremely toxic.

Lead poisoning is the number-one indoor environmental threat to children. High levels of lead exposure can cause permanent brain damage, coma, convulsions, and even death. Thankfully, these high levels are now relatively rare in the U.S. It's much more common for children to suffer chronic, low-level exposure. Low-level lead exposure can cause reduced IQ and attention span, hyperactivity, impaired growth, learning disabilities, and a range of other health issues. Lead is most harmful to children under six because it is easily absorbed into their growing bodies, where it interferes with the developing brain and other organs and systems. Although childhood lead poisoning in the U.S. has dramatically decreased in the past 20 years due to limits on lead in consumer products, lead poisoning is still a problem affecting an estimated 890,000 children under five. Pregnant women and women of childbearing age are also at increased risk, because lead ingested by the mother can cross the placenta and affect the fetus.

Exposure to lead-contaminated dust from deteriorated paint in poorly maintained older homes is the primary means of lead poisoning in children. Lead dust is invisible to the naked eye. It settles quickly and is very difficult to clean up. The dust gets on toys and hands and can then be ingested through normal hand-to-mouth activity. A smaller number of cases— often more serious ones—are caused by repainting and remodeling projects that disrupt old painted surfaces without proper precautions to contain and clean up the lead dust. Lead from sources like old gasoline or paint chips can also contaminate the soil around homes, where it can remain indefinitely.

Like asbestos, the mere presence of lead paint in a home is not necessarily dangerous. Even though an estimated two-thirds of all houses in this country contain some lead paint, the majority of children still live safely in these homes. If you purchase a home built before 1978, by law you should receive a federal government pamphlet dealing with lead hazards in a home, as well as any information the owner has on known lead paint hazards in the

If you purchase a home built before 1978, by law you should receive a federal government pamphlet dealing with lead hazards in a home as well as any information the owner has on known lead paint hazards in the home.

Deal Killers: Lead paint that is peeling in large quanties around a home might well be a deal breaker, especially if you have small children. Lead poisoning is regarded as the single most dangerous home health hazard there is. However, lead paint in good condition is less dangerous, especially once it is painted with a sealing coat of nonlead paint.

Many experts consider lead poisoning to be the number-one indoor environmental threat to children. High levels of lead exposure can cause permanent brain damage, coma, convulsions, and even death.

home. By law, you also have the right to have the property tested for lead hazards at your own expense before closing. If you have or plan to have children, it is advisable to have the home tested for lead paint hazards.

A thorough test for lead should evaluate each different painted surface. Different paints may have been used on doors, walls, window frames, and so on. Professional testers will use one of two methods to measure the lead in paint. One method uses portable x-ray devices. The other method involves laboratory testing of paint samples. The EPA does not recommend any of the numerous do-it-yourself lead tests. These kits are not sensitive enough to detect low levels of lead, so a sample might test negative but still have a hazardous lead content.

According to the experts, action should be taken to reduce exposure to lead when the lead content of paint exceeds 0.5 percent. Lead reduction measures are especially important when the paint is deteriorating, or when infants, young children, or pregnant women are present. Federally backed mortgages often require abatement or removal of peeling paint on houses built before 1978. Check with your mortgage lender if these requirements affect you.

If you are buying a house that contains lead paint, you have several options. Paint in good condition can be left alone. Damaged or peeling sections can be covered with drywall or some other building material. A third option is complete removal of lead paint in a house. This task should be done only by professionals trained in lead paint removal, because the process of removing paint can produce dust or fumes. In some cases, removal may involve complete replacement of items like windows, doors, and trim, because of the cost and difficulty of removing the paint. Before undertaking any abatement procedures, have a qualified lead inspector do a lead hazard risk analysis that considers all your options.

Lead abatement: $8 to $15 per square foot

To test for lead, dust samples are taken from different areas of the home and analyzed at qualified laboratories.

Formaldehyde

When I think of formaldehyde, I think of the smell of the mortuary where my uncle worked. You may be unaware that formaldehyde also is widely used in the manufacture of many building materials and household products. As a result, formaldehyde in the form of vapor can often be found in significant concentrations inside newer or newly remodeled homes. Exposure to formaldehyde can cause eye, nose, and throat irritation as well as coughing, rashes, headaches, and dizziness. Studies have also shown formaldehyde to cause cancer in lab animals, and it may cause cancer in humans. Symptoms of exposure to formaldehyde are often mistaken for a cold or allergies and generally go away once exposure is stopped.

Pressed wood products, especially those containing urea-formaldehyde glues, are one of the main sources of formaldehyde vapors in homes. These products include particleboard used in flooring underlayment, shelves, plywood wall panels, and the medium density fiberboard (MDF) used in drawers, cabinets, and furniture. While formaldehyde is still present in these products when they are new, manufacturers have reduced emissions from pressed wood products by 80 to 90 percent from the levels of the early 1980s.

A tightly sealed home can trap the relatively small amounts of formaldehyde emitted and lead to build-up over time. Adequate fresh-air ventilation eliminates the problem for most people.

Carbon Monoxide

Carbon monoxide (CO) is one of the most commonly encountered and deadly poisons found in homes, and the most easily preventable. Produced by woodstoves, kerosene heaters, fireplaces, furnaces, water heaters, gas stoves, and other fuel-burning appliances, this colorless and odorless gas is responsible for thousands of emergency room visits each year, as well as hundreds of unintentional deaths. Symptoms of CO poisoning resemble the initial stages of the flu, such as dizziness, nausea, and headache, but quickly go away when you go outside the home. Even at low levels, carbon monoxide leads to fatigue and chest pains in those with heart disease.

Preventing carbon monoxide poisoning is as easy as having combustion appliances checked annually, checking chimneys for

Carbon monoxide detectors are readily available in many styles.

proper caulking and intact mortar, and installing CO detectors on each floor. Detectors that run off current take samples on a recurring basis, while battery-powered models have passive sensors which react to prolonged exposure. Either type is safe for use, as long as it is UL (Underwriter Laboratories) approved.

Carbon monoxide detectors: $25 to $40

Drinking Water

The water in most homes is either supplied through public water utilities or is drawn from private wells. While public water is generally safe, homes that use well water have greater risks for possible contamination from bacteria, pesticides, fertilizers, and other pollutants that can leach into groundwater and contaminate wells. Contaminated drinking water is undetectable unless tested. If you plan to buy a home that is supplied by well water, contact the county environmental offices about any possible contamination problems in the area, such as farms, factories, or known spills. Keep an eye out for anything that might be seeping into the ground and ask about it. If the home now uses (or once used) oil for heating, for instance, there may be oil tanks buried on the lot, which can leak. If there is any reason for concern, have the water tested.

Municipal water can become contaminated once it enters the home. One of the leading dangers to a home's water supply lies in the pipes and fixtures of the home itself. Lead from old pipes, fixtures, or solder can dissolve in water and build to high concentrations when the water stands for several hours without use. Water which is especially soft (makes soapsuds easily) or slightly acidic can be very reactive with lead. These factors can contribute to elevated levels of lead in water. Ingested lead is a serious health danger to children and expectant mothers. Baby formula made from lead-contaminated tap water is a common cause of childhood lead poisoning. Lead in a mother's body can also transfer to children through nursing. In most areas, the use of lead solder in plumbing pipes has been banned since the 1980s.

To assess lead levels in a home's drinking water, you must have a water sample tested by an approved laboratory. As with lead paint, do-it-yourself kits may not measure low but still hazardous levels. When taking samples, most

A damaged furnace flue can leak carbon monoxide into living areas. This chimney either has a missing cap or a blocked drainage tube.

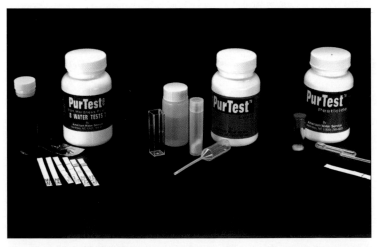

Do-it-yourself drinking water tests abound—but professional laboratory testing is the only sure way to get accurate results.

tests require a "first draw" sample, taken after the water has sat in the pipes for at least four hours, and a "fully purged" sample, after the water has run continuously for at least a minute. If you are taking the samples yourself, follow the sampling procedures precisely to ensure accurate results.

If your home water supply shows high lead concentrations, there are remedies beyond replumbing the house. Purging standing water from the pipes by running the water for one full minute before use can usually eliminate high lead concentrations temporarily.

Electromagnetic Fields

Suppose you find your dream home, but towering across the property are strings of high-voltage power lines. Besides being unsightly and noisy, will the electromagnetic fields generated by power transmission lines harm you or your family? Most of the concern about this issue began in the early 1990s, when small studies seemed to show possible connections between electromagnetic fields generated by power lines and cancer. Further studies have not been able to find a conclusive link between EMF exposure and cancer, though suspicion remains.

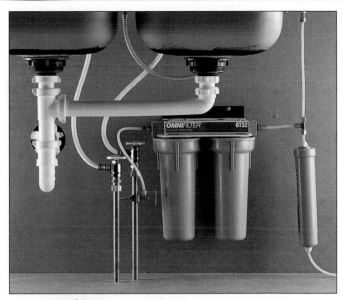

Signs of Gold:
A good-quality water filter has several filter elements to remove different impurities in potable water. This particular model has a carbon filter to remove odors, a sediment filter to remove large impurities and clear up cloudiness, and an additional filter leading to a refrigerator icemaker. If a water test indicates more serious pollutants, though, a more sophisticated reverse-osmosis filter would be advisable.

If your home water supply shows high lead concentrations, there are remedies beyond replumbing the house.

Power lines that create electromagnetic fields are not proven health hazards, though there is much debate on this subject.

Additional Concerns

The hazards listed in this chapter are the most common, but they are by no means the only environmental problems you may encounter in or around a house. All sorts of problems, from noise and light pollution to ash from annual forest fires to industrial pollution, can cause quality of life and health problems—not to mention cost.

Get to know the problems unique to your region by talking with homeowners' associations, university extensions, or even local newspaper web sites, which may have searchable back issues. And always ask the homeowner whenever you have concerns.

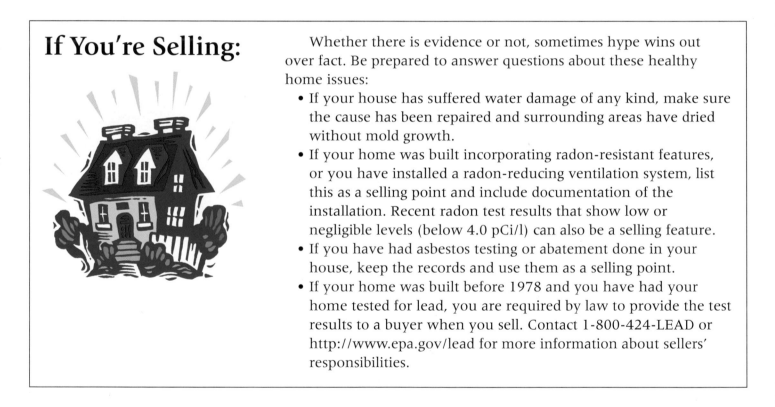

If You're Selling:

Whether there is evidence or not, sometimes hype wins out over fact. Be prepared to answer questions about these healthy home issues:

- If your house has suffered water damage of any kind, make sure the cause has been repaired and surrounding areas have dried without mold growth.
- If your home was built incorporating radon-resistant features, or you have installed a radon-reducing ventilation system, list this as a selling point and include documentation of the installation. Recent radon test results that show low or negligible levels (below 4.0 pCi/l) can also be a selling feature.
- If you have had asbestos testing or abatement done in your house, keep the records and use them as a selling point.
- If your home was built before 1978 and you have had your home tested for lead, you are required by law to provide the test results to a buyer when you sell. Contact 1-800-424-LEAD or http://www.epa.gov/lead for more information about sellers' responsibilities.

The Healthy Home: Avoiding the "Sick House" Checklist

Item	Good Indications	Condition		
		Good	Average	Poor
Indoor Air	• Air-to-air exchanger	☐	☐	☐
	• Air filtration system	☐	☐	☐
Moisture/ mold	• No history of flooding	☐	☐	☐
	• Leaks not present	☐	☐	☐
	• No visible mold	☐	☐	☐
Radon	• Radon remediation	☐	☐	☐
	• Radon test results	☐	☐	☐
Lead	• House newer than 1980	☐	☐	☐
	• Lead test results	☐	☐	☐
Drinking water	• Plumbing newer than 1980	☐	☐	☐
	• Well test results	☐	☐	☐
	• No neighboring hazards	☐	☐	☐
Carbon monoxide	• Chimney well maintained	☐	☐	☐
	• Furnace inspection record	☐	☐	☐

Chapter 18

Internet Resources

There are many resources available to home buyers on the Internet—more than we could ever list here. If you come across a house with an unfamiliar construction technique or material, or an unusual built-in accessory not covered in this book (a built-in vacuum system, perhaps, or hay bale framing), the Internet can be an especially good resource.

Unfortunately, many web sites devoted to products and building techniques are either biased (because their manufacturer created them) or not entirely authoritative (remember, nobody checks this stuff). It is always a good idea to look carefully at who is behind the information being posted on a web site you visit. Try to get a balanced viewpoint by checking several sites.

What follows represents a few of the available good sources of general information as well as some tips on sources for more specific information.

It is always a good idea to look carefully at who is behind the information being posted on a web site you visit. Try to get a balanced viewpoint by checking several sites.

The Elements of a House

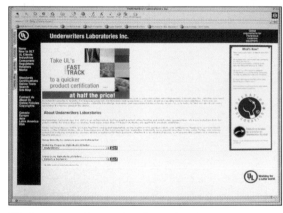

Visit www.ul.com (Underwriters' Laboratory) to find resources related to product safety in virtually every area of the home.

These sites are good places to start to get more information on the things you'll find in any house, how to evaluate them, and what you'll need to care for them. Many of these industry associations and manufacturers have extensive consumer information and product certification information—they can help you know what you're buying when you buy a house.

Flooring
- Carpet and Rug Institute: www.carpet-rug.com
- National Wood Flooring Association: www.woodfloors.org
- Resilient Floor Covering Institute: www.rfci.com

Walls
- Steel Framing Alliance: www.steelframingalliance.com

Insulation & HVAC
- The U.S. Department of Energy: www.energy.gov
- Owens Corning: www.owenscorning.com
- Air Conditioning Contractors of America: www.acca.org
- Air-Conditioning and Refrigeration Institute: www.ari.org

Windows
- Department of Energy: www.energy.gov
- Efficient Windows Collaborative: www.efficientwindows.org

Cabinets
- Kitchen Cabinet Manufacturers Association: www.kcma.org

Electrical
- National Fire Protection Association: www.nfpa.org
  This site includes the National Electrical Code (NEC) Digest in addition to excellent general information about fire safety.

Plumbing
- NSF International: www.nsf.org/plumbing

Siding
- Vinyl Siding Institute: www.vinylsiding.org
- Western Red Cedar Lumber Association: www.wrcla.org

Roofing
- National Roofing Contractors Association: www.nrca.net
- Asphalt Roofing Manufacturers Association: www.asphaltroofing.org
- Metal Roofing Alliance: www.metalroofing.com

General
- Underwriter Laboratories Inc.: www.ul.com
- National Association of Home Builders: www.nahb.com

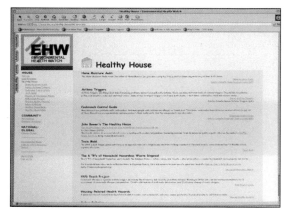

Check www.ehw.org (Environmental Health Watch) for information on home, community, and planetary environmental concerns.

Professional Inspectors

The American Society of Home Inspectors is the professional certification organization for home inspectors. Its web site features a national inspector directory, as well as a glossary useful for understanding inspection reports. The site also links you to specific state inspection laws.
- American Society of Home Inspectors: www.ashi.org

Finding & Financing a Home

The ABC's of Real Estate (a site run by a private California lender) and Homestore.com (the media company that operates the National Association of Realtors' web site) are both excellent resources, providing useful information and tools like mortgage calculators to help you understand how much you can afford, database search engines that give you recent sale prices in a neighborhood, and information on nearby schools and services.

A visit to Homestore.com (www.homestore.com) gives a wide variety of information on locating homes and obtaining financing.

This book cannot cover all aspects of healthy house research, and you may have specific health concerns that will factor into your purchasing decision. If so, you will want to do some additional research. Do so carefully.

- www.realestateabc.com
- www.homestore.com and www.realtor.com

Real Estate Professionals

If you need an agent, a number of organizations offer information on finding and working with real estate agencies across the country.

- International Real Estate Digest: www.ired.com
- National Association of Realtors: www.realtor.com

Federal Affordability Programs and Information

The U.S. Department of Housing and Urban Development (HUD) is the government agency concerned with affordable and low-income housing, and its web site provides information on agents who can sell such houses. HUD also provides information on a variety of federal home repair and mortgage lending programs, federal tax incentives and credits, and federal fair housing laws.

- www.hud.gov

Lending and Financing Information

Fannie Mae and Freddie Mac are private lenders specializing in low-interest loans for low-, moderate-, and middle-income home buyers. The Freddie Mac web site also features an excellent glossary of real estate and home lending terms.

- Fannie Mae (Federal National Mortgage Association): www.fanniemae.com
- Freddie Mac (Federal Home Loan Mortgage Corporation): www.freddiemac.com

Energy Efficiency & Healthy House Information

Government information clearinghouses and other federal resources are a good place to find all sorts of information, including tips on energy efficiency, energy usage codes, and weatherization, as well as information on government incentive programs.

- Consumer Energy Information: www.eere.energy.gov/consumerinfo
- Home Energy Saver: www.homeenergysaver.lbl.gov
- Environmental Protection Agency: www.epa.gov

A variety of not-for-profit, commercial, and private agencies offer information on energy efficiency and healthy houses for homeowners. As I said in the Healthy House chapter, some sick house issues are quite controversial, so it's *very* important to consider the source of any information.

This book cannot cover all aspects of healthy house research, and you may have specific health concerns that will factor into your purchasing decision. If so, you will want to do some additional research. Do so carefully. You'll find seemingly helpful web sites and magazine articles that will tell you that "toxic mold" is an endemic, imminent danger to families and others that will tell you that mold causes nothing more than allergies—it all depends on whether insurance providers or trial lawyers are behind the information. Make sure any concerns about mold—or any other potential hazards—in a prospective home are well founded before you factor them into your decision.

Energy Efficiency
- Alliance to Save Energy: www.ase.org
- Consortium for Energy Efficiency: www.cee1.org

Healthy House
- American Lung Association Health House Program: www.healthhouse.org
  The ALA website has information on dozens of potential indoor air pollutants and how to deal with them.
- Environmental Health Watch: www.ehw.org
- The Environmental Protection Agency: www.epa.gov
  This impartial government resource is a good, objective place to find information about potential mold issues as well as other home environment concerns.

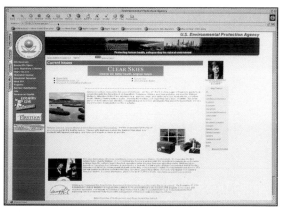

The U. S. Environmental Protection Agency (www.epa.gov) offers objective information on many areas of concern for homeowners.

Regional Information

University extensions are an excellent, unbiased source for all kinds of information, including regularly updated information on local and regional healthy house and energy efficiency issues, regional affordability and subsidy programs, and a variety of other useful information for homeowners. Most extensions have web sites. The following sites represent only a sample available at the time of this book's publication. Check with an area state university or your favorite search engine to find additional similar resources in your area.
- University of Colorado: www.ext.colostate.edu
- University of Illinois: www.extension.uiuc.edu
- Iowa State University: www.extension.iastate.edu
- North Carolina State University: www.ces.ncsu.edu
- Michigan State University: www.msue.msu.edu
- University of Minnesota: www.extension.umn.edu
- Washington State University: www.energy.wsu.edu
- University of Wisconsin: www.uwex.edu

Index

CONTRIBUTORS

ACT, Inc. Metlund-Systems
800-638-5863
info@gothotwater.com

The Bilco Company
203-934-6363
www.bilco.com

GE Consumer Products
1-800-626-2000
www.geappliances.com

JELD-WEN, Inc.
800-877-9482
www.jeld-wen.com

Kolbe and Kolbe Millwork Co., Inc.
800-955-8177
www.kolbe-kolbe.com

University of Nebraska Department
    of Entomology
402-472-8691
http://entomology.unl.edu

Weil-McLain
219-879-6561
www.weil-mclain

PHOTOGRAPHERS

Karen Melvin
Minneapolis, MN
©Karen Melvin for Sala Architects
    p. 64

Robert Perron
Branford, CT.
© Robert Perron p. 65

R Todd David Photography
St. Louis, MO
© R. Todd Davis p. 35

Index Stock Imagery, Inc.
New York, NY
© Index Stock Imagery, Inc./
    Philip Wegener-Kantor
        P. 81 (top)

Bill Tijerina Photography
Columbus, Ohio
©Bill Tijerina p. 84